A Naturalist's Guide to
Seashore Plants
An Ecology for Eastern North America

Donald D. Cox
Illustrations by Shirley A. Peron

Syracuse University Press

Copyright © 2003 by Syracuse University Press
Syracuse, New York 13244-5160

All Rights Reserved

First Edition 2003

03 04 05 06 07 08   6 5 4 3 2 1

The paper used in this publication meets the minimum requirements of American National Standard for Information Sciences—Permanence of Paper for Printed Library Materials, ANSI Z39.48–1984∞™

Library of Congress Cataloging-in-Publication Data

Cox, Donald D.
A naturalist's guide to seashore plants : an ecology for eastern North America / Donald D. Cox ; illustrations by Shirley A. Peron.— 1st ed.
p. cm.
Includes bibliographical references (p.).
ISBN 0–8156–0778–4 (pbk. : alk. paper)
1. Seashore plants—Atlantic Coast (North America)—Ecology. 2. Seashore plants—Gulf Coast (U.S.)—Ecology. 3. Seashore ecology—Atlantic Coast (North America) 4. Seashore ecology—Gulf Coast (U.S.) I. Title.
QK110 .C68 2003
581.7'699—dc21
200215162

Manufactured in the United States of America

A Naturalist's Guide to Seashore Plants

To the next generation
Cindy, Kurt, Olivia, Lisa, Mark, Allison, David, *and* Big John

Donald D. Cox studied at Marshall University, received a Ph.D. from Syracuse University, and was a professor of biology for forty-one years. His publications include *Some Postglacial Forests in Central and Western New York State; The Context of Biological Education: The Case for Change; Common Flowering Plants of the Northeast; Seaway Trail Wildguide to Natural History;* and from Syracuse University Press, *A Naturalist's Guide to Wetland Plants: An Ecology for Eastern North America.*

Contents

Illustrations ix
Acknowledgments xiii
Introduction xv

1. The Ocean and the Seashore 1
2. Types of Plants 27
3. Adaptations for Survival 44
4. Sand Dunes and Beaches 54
5. Salt Marshes 66
6. Through the Year 74
7. Plants of Special Interest 92
8. Naming, Collecting, and Preserving Plants 109
9. Activities and Investigations 130

Glossary 141
Bibliography and Further Reading 145
Index 151

Illustrations

1.1. Cross section of leaf   5
1.2. Gulfweed *(Sargassum fluitans)*   9
1.3. Diatoms   10
1.4. Broad-leafed Kelp *(Laminaria saccharina)*   19
1.5. Eelgrass *(Zostera marina)*   21
1.6. Widgeon grass *(Ruppia maritima)*   22
1.7. Turtlegrass *(Thalassia testudinum)*   22
1.8. Prop roots   24
1.9. Red Mangrove *(Rhizophora mangle)*   24
1.10. Black Mangrove *(Avicennia germinans)*   25
1.11. White Mangrove *(Laguncularia racemosa)*   25
2.1. Sea Lettuce *(Ulva lactuca)*   29
2.2. Rockweed *(Fucus vesiculosus)*   29
2.3. Dulse *(Palmaria palmata)*   30
2.4. Sac Fungus   31
2.5. Gill Mushroom   32
2.6. Bracket Fungi   33
2.7. Rough-Stemmed Boletus *(Boletus scaber)*   33
2.8. Earthstar *(Geastrum saccatum)*   34
2.9. Crustose Lichen, Foliose Lichen, and Fruticose Lichen   35
2.10. Mosquito Fern *(Azolla caroliniana)*   37
2.11. Lace-Frond Grape-Fern *(Botrychium dissectum)*   38
2.12. Bald Cypress *(Taxodium distichum)*   39
2.13. Atlantic White Cedar *(Chamaecyparis thyoides)*   39
2.14. Angiosperm flower   41

2.15. Perennial Salt-Marsh Aster *(Aster tenuifolius)*   42
3.1. Prickly Pear *(Opuntia humifusa)*   45
3.2. Lance-Leaved Milkweed *(Asclepias lanceolata)*   47
3.3. Japanese Rose *(Rosa rugosa)*   47
3.4. Swamp Rose-Mallow *(Hibiscus moscheutos)*   48
3.5. Type 1 flower; type 2 flower   49
3.6. Narrow-Leaved Cattail *(Typha angustifolia)*   49
3.7. Beach-Plum *(Prunus maritima)*   51
3.8. Cocklebur *(Xanthium strumarium)*   52
3.9. Tall Wormwood *(Artemisia campestris)*   53
3.10. Pond Pine *(Pinus serotina)*   53
4.1. Beach-Grass *(Ammophila breviligulata)*   55
4.2. Sea-Oats *(Uniola paniculata)*   56
4.3. Seaside Spurge *(Euphorbia polygonifolia)*   57
4.4. Sea-Rocket *(Cakile edentula)*   58
4.5. Sea-Purslane *(Sesuvium maritimum)*   58
4.6. Sea-Elder *(Iva imbricata)*   59
4.7. Beach Pea *(Lathyrus maritimus)*   59
4.8. Beach Wormwood *(Artemisia stelleriana)*   60
4.9. False Heather *(Hudsonia tomentosa)*   60
4.10. Sea-Beach Sandwort *(Honckenya peploides)*   60
4.11. Seaside-Goldenrod *(Solidago sempervirens)*   60
4.12. Dune Sandspur *(Cenchrus tribuloides)*   61
4.13. Groundsel Tree *(Baccharis halimifolia)*   61
4.14. Croton *(Croton punctatus)*   62
4.15. Pennywort *(Hydrocotyle bonariensis)*   62
4.16. Seaside Evening-Primrose *(Oenothera humifusa)*   62
4.17. Yaupon *(Ilex vomitoria)*   62
4.18. Bayberry *(Myrica pensylvanica)*   63
4.19. Live Oak *(Quercus virginiana)*   64
4.20. Pitch Pine *(Pinus rigida)*   64
5.1. Smooth Cord-Grass *(Spartina alterniflora)*   67
5.2. Salt-Meadow Grass *(Spartina patens)*   68
5.3. Spike-Grass *(Distichlis spicata)*   68
5.4. Marsh Elder *(Iva frutescens)*   68
5.5. Perennial Glasswort *(Salicornia virginica)*   68
5.6. Samphire *(Salicornia europaea)*   70

5.7. Southern Sea-Blite *(Suaeda linearis)*   70
5.8. Salt-Marsh Sand-Spurry *(Spergularia marina)*   70
5.9. Salt-Marsh Agalinis *(Agalinis maritima)*   70
5.10. Coast-Blite *(Chenopodium rubrum)*   71
5.11. Spearscale *(Atriplex patula)*   71
5.12. Sea Oxeye *(Borrichia frutescens)*   72
5.13. Sea Lavender *(Limonium carolinianum)*   72
5.14. Common Reed *(Phragmites australis)*   73
5.15. Seaside Plantain *(Plantago maritima)*   73
6.1. Broom Crowberry *(Corema conradii)*   77
6.2. Eastern Blue-Eyed Grass *(Sisyrinchium fuscatum)*   78
6.3. Seaside Knotweed *(Polygonum glaucum)*   80
6.4. Yellow Hedge-Hyssop *(Gratiola aurea)*   81
6.5. Saltwort *(Salsola kali)*   81
6.6. Flat-Topped Goldenrod *(Euthamia tenuifolia)*   83
6.7. Jointweed *(Polygonella articulata)*   83
6.8. Seabeach Orache *(Atriplex arenaria)*   84
6.9. Salt-Marsh Fleabane *(Pluchea odorata)*   89
7.1. Scrub Oak *(Quercus ilicifolia)*   94
7.2. Bracken Fern *(Pteridium aquilinum)*   98
7.3. Loblolly Pine *(Pinus taeda)*   99
7.4. Poison Ivy *(Toxicodendron radicans)*   99
7.5. Horse-Mint *(Monarda punctata)*   102
7.6. Horseweed *(Conyza canadensis)*   103
7.7. Irish moss *(Chondrus crispus)*   106
7.8. Winged Kelp *(Alaria esculenta)*   107
9.1. Leaves and stems of (1) grasses, (2) sedges, and (3) rushes   137

Acknowledgments

I am indebted to Barbara Cox and Shirley Peron for providing many helpful suggestions. I especially wish to thank Sharon Doerr for carefully reading the manuscript and offering suggestions for improvement. I am grateful for the courtesy and cooperation of the staff of Penfield Library at SUNY-Oswego. To the staff at Syracuse University Press I extend my thanks for creative suggestions and professional guidance.

For the botanical and common names of plants I have, when applicable, used those presented in the second edition (1991) of the *Vascular Plants of Northeastern United States and Adjacent Canada* by Henry A. Gleason and Arthur Cronquist. For plants growing outside the range of the above work, I have relied on the *Manual of Southeastern Flora* by John Kunkle Small.

Introduction

Ecology is the study of organisms in relation to their environments. This book is about the ecology of the plants that grow along the Atlantic and Gulf coasts of North America. It includes not only those that are ordinarily called land plants but also the microscopic ones that inhabit the oceans and those that grow in the relatively shallow water on the shoreward margins of the continental shelf. Each organism has a unique set of features that enable it to survive in its environment. Each is a vital part of the ecosystem, and the ocean and seashore ecosystems are essential components of the ecology of the earth.

In the following chapters, for ease of reading, technical terminology has been kept to a minimum. Some terms used to describe plants, although not highly technical, may have special meanings that are unfamiliar to the reader. For these, a glossary has been included. As one's interest in and knowledge of plants grows, invariably a point is reached where common names are no longer satisfactory. Relying on common names can be confusing because every region may have a different name for the same plant. For this reason, the first time a plant's common name is used in each chapter, the botanical name is given in parenthesis. With a little practice, these will become as familiar as common names and they are much more reliable. The botanical name for a species is the same all over the world.

The way oceans and seashores function as ecosystems and their role in the ecology of the earth are discussed in chapter 1. In chapter 2, brief descriptions are given for the major types of plant groups that grow in seashore habitats, with a focus on the flowering plants or angiosperms. Emphasis has been placed on field observations, the intention being to provide descriptions and drawings that will aid in identification. With a little effort an alert observer can learn to recognize the specific characteristics of many species. In a given region of North America there may be several thousand species of plants. Trying to learn all of these is a daunting challenge and the beginner may wish to begin by learning the plants in a limited group. For example, he or she could start by learning the green, brown, and red macroscopic algae, the plants of beaches and dunes, or the species in salt marshes. All of these groups and others are described in the following chapters.

Seashore plants have special characteristics that enable them to survive in the demanding coastal habitat. These are described in chapter 3. Chapters 4 and 5 describe the ecology of sand dunes and salt marshes and the main plants that inhabit them. Chapter 6 describes the changes that take place through the seasons. Chapter 7 presents toxic, medicinal, and edible plants of seashore habitats. It includes a section on poisonous algae and the red tide. Chapter 8 details inexpensive methods for those interested in collecting and preserving plants. Chapter 9 is offered for those who wish to go a step beyond identifying or collecting plants. It includes activities, projects, and investigations.

Observing and learning about the seashore can be entertaining and educational. Since plants cannot run away, they can be studied in detail in their natural habitats. While this is convenient for all who study or observe them, it also makes plants vulnerable to all sorts of destructive forces. Their greatest threat is disruption of habitats resulting from human activities. The destruction of natural areas is increasing as the human population grows, making habitat conservation an urgent priority.

The ability to identify plants has its own reward in personal satisfaction. Recognizing some of the plant species when visiting a different seashore is like seeing old friends. It is comforting to know that even in widely separated environments there are familiar "faces." In addition, as one travels along the seashore, being aware of the communities that contribute to the landscape gives that landscape a richer meaning and enhances the enjoyment with which it is viewed. It is the aim of this book to give the reader a broader understanding and a greater appreciation for the plants of seashore ecosystems.

Three additional publications are available in the "Naturalist's Guides" series. These are *A Naturalist's Guide to Forest Plants*, *A Naturalist's Guide to*

Meadow Plants, and *A Naturalist's Guide to Wetland Plants.* Topics covered in these publications include plant lore, ecology, and tips for plant identification; poisonous, hallucinogenic, medicinal, and food plants; and collecting and preserving plants. In addition there are activities, projects, investigations, and thought stimulators. For naturalists and other lovers of the outdoors, these books not only provide background to an ecosystem, but they can also serve as reliable field guides.

A Naturalist's Guide to Seashore Plants

1

The Ocean and the Seashore

Origins of the Oceans

A commonly accepted hypothesis for the origin of earth holds that the bodies in the solar system condensed from a rotating nebular cloud of gases and particulate matter. As this cloud collapsed under the influence of gravity, it contracted, began to spin faster, and flattened. Most of the matter was concentrated in the center where heat from the increasing internal pressure eventually initiated nuclear fusion to form the sun. The flat disc of matter spinning around the sun supplied the material that formed into planets.

The earth is thought to have formed in a similar manner but on a smaller scale than that which formed the sun. Matter accumulated around a core that attracted still more matter as its mass and gravity increased. As the earth grew, it was probably rocked by countless collisions with comets and meteors. One collision was so massive that the heat generated may have caused the entire early earth to become liquefied, with a great wave thrown into space to form the moon. As the earth continued to grow in mass, the internal pressure increased with a corresponding increase in temperature. There is evidence that the earth eventually became molten. This allowed heavy elements to concentrate in the earth's core and lighter ones to form the exterior layers.

The first atmosphere probably consisted of hydrogen and helium that was present in the original nebular cloud. With the onset of nuclear fusion in the sun, a solar wind of light and ionized particles essentially blew the helium and hydrogen away. Most investigators believe that volcanic eruptions were common on the surface of early earth. These and frequent meteor collisions would have contributed gases to an atmosphere that was quite different from the one that envelops the earth today. It would have contained large quantities of water vapor, carbon dioxide, and nitrogen. Iron oxide in the oldest rocks indicate that free oxygen was also present.

When the earth and the atmosphere cooled sufficiently, the water vapor condensed and it rained, filling the depressions on the irregular surface of the earth. This was a rain the likes of which the earth has not seen since and will never see again. The rain was hot, the oceans were near boiling, and it rained for thousands of years. A still unanswered question is where all of this water came from. Some investigators think it came from gaseous water within the surface layers of the earth; others believe much of the ocean water came from meteor and comet collisions.

The oceans today consist of about 95 percent water, by weight, and 3.5 percent dissolved solids, mostly sodium chloride or salt. This is the salinity, and it is usually expressed as grams of salt per kilogram of sea water. The average ocean salinity is 3.5 percent. As a result of the constant and thorough mixing by ocean currents and wave action, the salinity is fairly constant throughout the oceans of the world. Investigators are not in agreement as to the source of the salt, but all agree that the salinity has been the same for most, if not all, of the history of the oceans. Consequently, most of the evolution of living things occurred in salt water.

This description of the origin of the earth and the oceans is very different from the one presented by Rachel Carson in her remarkable 1951 book *The Sea Around Us*. Her account was based on current scientific knowledge of her time. Views are different today as a result of five decades of observation, investigation, and technology. The views of these events in 2050 will no doubt be different from those of today.

The Origin of Life

Life on earth is very old. The oldest objects that can be interpreted as fossils of living things are about 3.5 billion years old. To reach the stage where organisms would make a preservable impression in rocks required countless millions of years of evolution. The earth is believed to be 4.5 to 5 billion years old. It is likely life began sometime between 3.5 and 4 billion years ago.

There are no certainties with regard to the origin of life, but a reasonable expectation is that it originated in water. This is supported by the observation that water is the main component in living things today. It is the medium that supports all of the physiological processes that sustain living activities. Protoplasm, the living substance of cells, consists of 90 percent or more of water.

The earth is unique in the solar system in being at just the right distance from the sun for water to exist in the liquid state. The inner planets of Venus and Mercury are too hot for liquid water to exist, and the outer planets, with the exception of Mars, are too cold. Mars probably contained liquid water in its early years, and maybe even primitive life, when its active volcanoes were producing the greenhouse gas carbon dioxide in great quantities. But Mars is much smaller than Earth and farther from the sun; it cooled quickly when its volcanoes died and no longer produced carbon dioxide.

Most scientists accept the current view that life originated on Earth through chemical evolution. There are differences of opinion as to the sources of amino acids, which are the raw materials for this process. At one time, it was thought that the atmosphere contained large amounts of water vapor, methane, and ammonia. Lightning flashing through this mixture provided energy for the synthesis of amino acids and other carbon compounds. These rained into the oceans where molecular evolution proceeded. The presumed atmospheric conditions in this scenario have been created in a closed laboratory system, using electrical sparks as lightning. The results were the production of amino acids as predicted by the theory.

Another theory proposes that hydrogen cyanide and formaldehyde in the early atmosphere, both deadly poisonous to living cells, reacted with water to form amino acids. In the laboratory they also react to produce the chemicals that are essential for the formation of the spiral DNA molecule. These compounds are common in comets that have been studied. They could have reached the early earth through numerous collisions with comets and comet fragments.

Hot springs on the ocean floor offer another possibility for the chemical evolution of life. They occur in all the major oceans of the earth along fractures in the deep ocean floor. These may have been more active 4 billion years ago than they are today. The hot waters circulating through deep fractures were rich in dissolved chemicals just as they are today. They could have provided the raw materials, and heat from the fractures the energy, for the reactions that would have formed the building blocks of life.

Whatever the source of the building blocks, after countless millions of years, a self-duplicating DNA molecule and its attendant protoplasm evolved to a stage that could be called a living thing. No one knows what this primordial organism looked like because it was much too fragile to leave a fossil impression in the sediments that became rocks. It would have been classified as an animal rather than a plant because it did not manufacture its own food. Its source of food and energy for duplicating itself were the same compounds from which it had formed. The evolution of life on earth was a one-time-only event because after living things were present, the environment was so changed that the process could never be repeated. By the time living things developed a structure firm enough to leave a fossil record, life was well on its way to the great diversity we see today.

The next major event in earth history was the development of organisms that could manufacture their own food. The first of these were organisms somewhat like modern-day bacteria. Both the bacteria and blue-green algae of today have retained the characteristics of these organisms. Because they do not have a nucleus with their genetic material (DNA) enclosed in a membrane, they are classified as a separate kingdom of organisms, the Monera.

After many millions of years of evolution, the green algae arose from blue-green ancestors. Both the green and the blue-green algae added oxygen to the atmosphere through the process of photosynthesis. After more than a billion years, enough oxygen had been added to the atmosphere to form an ozone layer several miles above the earth's surface. This greatly reduced ultraviolet radiation so that 450 to 500 million years ago, certain green algae could make the transition from water, and the first land plants appeared. Until that ozone layer was thick enough, all life had been confined to water where it was protected from deadly cosmic radiation.

The green algae differed from their blue-green predecessors in having a nucleus with genetic material enclosed in a membrane. All land plants evolved from the green algae and possess this trait. The origin of green algae thus marks the origin of plant life on earth.

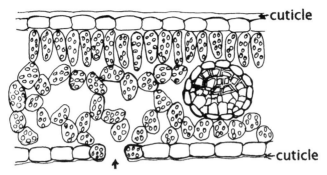
1.1. Cross section of leaf

The Origin of Life on Land

The most radical change that ever happened to plant life on earth was the transition from water to a land environment. The first land plants evolved from green algae living on the ocean shore in the area between low and high tide, the intertidal zone, where they had daily periods of exposure to the air. This was a time when geological forces within the earth were causing the region to be pushed upward. As the intertidal zone was elevated, it first became an area of tidal pools, then, after further lifting, even these disappeared. During the millions of years required for this process, those species whose rate of evolution kept pace with the rate of shoreline elevation became land plants. Many species that were unable to make the transition to land became extinct.

The fossil record indicates that the first land plants may have resembled low growing, branching green sticks. Later the ends of branches flattened and fused to form leaves. The plants that survived the rigors of natural selection had several structures that are necessary for survival on land. One was a root system that anchored the plant to the soil and absorbed mineral nutrients. Another was a waterproof, waxy covering, the cuticle, on aboveground parts, which prevented extinction from excessive water loss. A third feature was a series of tiny closable openings in the cuticle, the stomata, that allowed an exchange of carbon dioxide and oxygen during photosynthesis (fig. 1.1). Finally, survival on land required a conductive system to transport water and mineral nutrients from the roots upward and food material from photosynthetic tissue downward to the roots.

The Oceans Today

The Extent of the Oceans

The earth is truly a water planet. Only 29 percent of its total area is dry land. The other 71 percent is covered by oceans of varying depths, but the depths by far exceed the continental heights. If the earth was perfectly smooth, it would be covered by water to a depth of 8,953 feet (2,686 m). The oceans consist of three major branches extending northward from the Antarctic Ocean: the Atlantic is the longest, extending from the Antarctic to the Arctic Ocean; the Pacific is the largest, covering about one-third of the earth's surface; and the Indian Ocean is the shortest, covering about one-seventh of the earth's surface and lying mostly in the tropics.

The total volume of water in the oceans is enormous: millions of cubic miles or over a billion cubic kilometers. Another and perhaps more understandable way of viewing this is by considering the rate of turnover or renewal. The water in the oceans is completely renewed every forty thousand years. This means that at the current rate of stream flow it will require forty thousand years to equal the volume of water in the oceans.

The oceans exist in basins separated by continental land masses. The submerged borders of the continents slope gently seaward before they dip sharply into the ocean depths. These submerged borders are called continental shelves. Their widths are about 40 miles (65 km) and they average about 430 feet (129 m) in depth on the outer edge. They make up only about 7.5 percent of the total area of the sea floor. But as a result of the abundance of nutrients from inflowing streams, they have a greater concentration of living things than any other area on the ocean floor.

The Tides

A characteristic of the oceans that has profound effects on seashore life is the tidal fluctuations in water level. The tides are caused by a bulge in the waters of the oceans formed by the gravitational attraction between the earth, the sun, and the moon. When these are in line, as at full moon and new moon, the gravitational pull is greatest and the bulges largest. At this time the tides are highest and are called spring tides. When a line from the sun to the earth forms a right angle with one from the moon to the earth, as at first- and third-quarter moons, the tides are lowest and are called neap tides. There are two spring tides and two neap tides with each revolution of the moon around the earth. Thus, spring tides can be expected every fourteen days, followed seven days later by neap tides.

As the earth rotates on its axis, the bulge in the water of the oceans stays in the same relative position with regard to the moon. The result is two high tides and two low tides daily. This is typical of the shoreline along the eastern coast of North America. The orbit of the earth around the sun is not a perfect circle; the earth is closer to the sun in January and farther from the sun in July. Consequently, the gravitational pull of the sun is greatest, and the spring tides highest, in January. Conversely, the pull is least in July and the spring tides are lowest.

The Wind Belts of the Earth

Intense heating by the sun in the equatorial region and the earth's rotation have resulted in the development of wind belts on the surface of the earth. Warmer, less dense air in the equatorial zone rises, creating a low pressure area and daily showers. In the upper layers of the atmosphere, the rising air drifts northward, and then begins to settle back to the earth around 30 degrees north latitude. The same pattern occurs in the southern hemisphere, but in this discussion we will consider only the northern hemisphere. As the air descends, it warms, the relative humidity drops, and a dry air/ high pressure area is produced. This zone is associated with some of the driest regions on earth such as the Sahara Desert and the deserts of southwestern North America.

Both the equatorial low pressure zone and the high pressure area at 30 degrees north latitude are characterized by very little horizontal air movement. Sailors called the equatorial zone the doldrums because sailing ships could make little headway. The high pressure region is called the horse latitudes because, according to legend, Spanish sailors threw horses overboard because there was not enough water on the becalmed ships to support both sailors and animals.

The air settling to the earth in the horse latitudes spreads northward and southward. That moving to the south is deflected to the west by the earth's rotation. These are the trade winds, named for their importance in the days of sailing ships. The air that spreads northward is deflected to the east by the earth's rotation to become the prevailing westerlies. They cover an area between approximately 30 and 60 degrees north latitude. This includes most of the middle section of North America.

The wind belts are very important components of the water cycle. Radiant energy from the sun enhances evaporation from the oceans. This supplies water vapor for most of the precipitation that falls on earth. The wind belts transport the water-laden air to continental land masses where it eventually encounters cooling conditions and falls to the earth as rain or snow. The water circulates over the continents and is finally returned to the oceans in rivers and streams, completing the cycle.

In North America, precipitation has four main sources: cold, moist air from the northern Pacific Ocean; tropical, moist air from the southeastern Pacific; cold, moist air from the North Atlantic Ocean; and tropical, moist air from the Gulf of Mexico. In the United States and southern Canada, water vapor from these sources generates precipitation in weather disturbances that move from west to east in the prevailing westerlies. Frequently atmospheric low pressure weather disturbances over the North Atlantic called "nor'easters" bring moisture to northeastern North America.

Ocean Currents and Weather

When a steady wind blows across the oceans, the surface layer of water is set in motion. The wind belts have thus created ocean currents in all the oceans of the earth. This can be seen in the Atlantic Ocean north of the equator. The westerly blowing trade winds push the water to the west forming the North Equatorial Current. Some of the water enters the Gulf of Mexico where it circulates and reemerges as the Florida Current, between Florida and Cuba. The Florida Current can be considered the beginning of the Gulf Stream that flows northward.

At Cape Hatteras, North Carolina, the Gulf Stream begins to be deflected to the northeast by the earth's rotation. At about the latitude of Newfoundland, the easterly flowing water becomes the North Atlantic Current. This flow bends southward and passes between Spain and the Azore Islands as the Canary Current. Off the coast of North Africa, the Canary Current is pushed westward by the trade winds and becomes the North Equatorial Current. This completes a great circular flow in the Atlantic Ocean.

The Gulf Stream is of particular interest since it flows along the eastern coast of the United States. It flows like a great river but greater in size or volume than any terrestrial river. East of Miami, Florida, the Gulf Stream is about 20 miles (30 km) wide and nearly 1,000 feet (300 m) deep. This is a volume of flow more than one hundred times greater than that of the Mississippi River. The rate of flow at the latitude of Miami is over 4 miles (5.5 km) per hour. By the time it reaches Cape Hatteras the current has decreased to slightly less than 3 miles (4.75 km) per hour.

The temperature of the Gulf Stream off the eastern shore of the United States varies from 77 to 82°F (25–28°C). Although the temperature decreases as the water flows eastward, the water carried by the North Atlantic Current is still relatively warm when it reaches the shores of England and western Europe. These countries are in the latitudes that in North America include Newfoundland and northern Canada. It is the warm Gulf Stream water that gives them a temperate rather than an arctic climate.

The Sargasso Sea is an area about the size of the United States located in the Atlantic Ocean roughly between Bermuda and the Azore Islands. It is surrounded by the currents that circle the Atlantic north of the equator. It lies in the high pressure horse latitudes where winds are light with correspondingly low water movement. A high evaporation rate into the dry air results in a higher salinity in this part of the ocean than in surrounding waters. The very low nutrient level in the water has led some scientists to refer to this area as a wet desert. In spite of this characterization, there are great quantities of brown algae that sometimes wash ashore and are called gulfweeds. The genus of this alga is *Sargassum*, for which the area is named. *Sargassum fluitans* (fig. 1.2) is one of two free-floating species that make up 90 percent of the bulk of the gulfweed in the Sargasso Sea.

Some scientists think that the most important function of the oceans is their regulation of global climate. In addition to supplying water vapor for the rain, snow, hail, and sleet that falls on Earth, the oceans also act as a global thermostat. The earth would experience extremely harsh maximum and mini-

1.2. Gulfweed (*Sargassum fluitans*)

mum temperatures if it were not for the moderating effect of the oceans. One of the ways that heat is distributed globally is by ocean currents. Warm water from the tropics flows into northern zones, as in the Gulf Stream, and cold water from polar regions flows into southern areas, as in the Labrador Current between Greenland and North America.

Another way the oceans exert a moderating influence on global temperatures results from the high heat-holding capacity of water. A large quantity of heat must be lost before the temperature of water will drop and a large quantity must be gained before the temperature will rise. The amount of heat required to bring about an increase in temperature is called the specific heat, and water has the highest specific heat of any known liquid. It has been calculated that the heat liberated in cooling 1 cubic meter of water (a cube 39 inches on a side) by 1 degree Celsius (1.8°F) is enough to raise the temperature of 3,000 cubic meters of air by the same amount. This capacity of water and the ocean currents softens the extremes caused by unequal heating of the earth by the sun.

Life in the Oceans

Phytoplankton. Probably the most numerous living things on earth are in a group known as plankton. These are microscopic organisms that inhabit the great expanses of all the oceans. There are two kinds of plankton: phytoplankton and zooplankton. Phytoplankton are one-celled plants that carry on photosynthesis. They may be responsible for more than half the plant growth in the oceans. Tiny animals that depend on phytoplankton for food are called zooplankton.

The most abundant plants of the phytoplankton are diatoms (fig. 1.3). A characteristic feature of diatoms is that they have cell walls of silica, the main component of glass; they can be said to live in glass boxes. Since they have no organs of locomotion to keep them near the surface, they run the risk of sinking below the level to which light can penetrate. Two adaptations help them to avoid this fate: they are so small that their rate of descent is very slow, and their food reserve is a drop of oil, which gives them added buoyancy. In addition, during the daylight hours when pho-

1.3. Diatoms

tosynthesis is taking place, the presence of gaseous oxygen within the cells causes the phytoplankton to be closer to the surface. At night they sink to deeper levels.

Diatoms that are not consumed by other organisms eventually die and sink to the bottom. Thick deposits of the transparent cell walls in areas that have been uplifted but were at one time on the ocean floor indicate that the process has been going on for hundreds of millions of years. These deposits are called diatomaceous earth or Fuller's earth, and at one site in California they are at least 1,000 feet (300 m) thick. Diatomaceous earth is very useful economically as a filtering agent in the sugar industry and as a fine abrasive agent in polishes such as silver polish and, at one time, toothpaste. It is also associated with petroleum, which may have originated from the oil droplet in each diatom cell. Some oil fields are overlain by thousands of feet of diatomaceous earth deposits.

Diatoms are appropriately called "the grass of the sea" because they are as numerous as blades of grass and are at a similar place in the food chain. It has been calculated that a cubic foot (0.028 cubic meters) of water may contain 72 million diatoms. In some regions this figure may be even larger. They are at the base of all food chains and webs in the ocean. For example, the largest creature on earth, the blue whale, lives on small shrimplike crustaceans called krill. Krill feed directly on plankton. Since there is a 90 percent loss when energy is transformed from one organism to another, it requires 22,000 pounds (10,000 kg) of plankton to produce 2,200 pounds (1,000 kg) of krill, which becomes only 220 pounds (100 kg) of whale tissue.

Some food webs are much more complex. Killer whales eat seals that consume fish and squid. The latter feed on krill that eat phytoplankton. In this instance, 22,000 pounds (10,000 kg) of krill produce less than $^1/_2$ pound (1 kg) of killer whale tissue. These examples illustrate the basic ecological principal that the longer the food chain, the less energy is available at the top.

Oxygen. The importance of phytoplankton extends far beyond oceanic food webs. Ancestral phytoplankton generated the oxygen in earth's early atmosphere that made life on land possible. Today, they are the main producers that maintain the 21 percent oxygen content of the atmosphere. It is through the photosynthesis of modern phytoplankton that life on land continues. One environmental interaction shared by humans and all mammals and probably the one most often taken for granted is breathing. If one of life's essential requirements could be called the most important, it is the oxygen in the air. Humans can survive for approximately a month without food and about a week without water but only a few minutes without oxygen.

Organisms that require oxygen use it to convert the chemical energy of food into energy necessary to live, grow, and move. Although they do not usually move the way animals do, most plants grow throughout their lives and they require energy to produce leaves, flowers, fruits, and seeds. The process of converting food into energy is called cellular respiration, and it takes place in all living things. Carbon dioxide is a by-product of cellular respiration. During daylight hours, plants carry on photosynthesis, which generates more oxygen than they use in respiration. The surplus diffuses into the environment. In addition to that released in respiration, most plants, including phytoplankton, must absorb additional carbon dioxide from their environment. In the dark, plants do not generate oxygen and like most other living things they release carbon dioxide.

Carbon Dioxide. Carbon dioxide is removed from the atmosphere by photosynthesis and returned to it by cellular respiration and decay. These processes balance each other and have kept carbon dioxide at about the same concentration for thousands of years. It currently makes up about .03 percent of the atmosphere, but it has not always been at this level. During past ages there may have been several thousand times more carbon dioxide in the air than today. It was reduced to the present concentration by photosynthesis in phytoplankton and in the coal age forests that were converted into deposits of petroleum and coal. These are the fossil fuels that provide 90 percent of the energy used by humans today.

The atmospheric carbon dioxide/oxygen balance maintained by photosynthesis and respiration/decay began to change with the invention of the steam engine and the initiation of the Industrial Revolution about 1800. Steam-operated machines using coal for fuel began to add more carbon dioxide to the air than was being removed by photosynthesis. At first the accumulation was very slight, but it increased as the use of coal, oil, and natural gas escalated. The invention and use of the internal combustion engine resulted in a great boost in emissions. Since the early 1800s, the concentration of carbon dioxide in the air has increased by 25 percent. The greatest increase has been in the last fifty years. Since 1958 its concentration has increased by 10 percent. It is now at a level estimated to be the highest in 130,000 years.

A change in world climate is almost a sure thing if the atmosphere continues to accumulate carbon dioxide. It is called a greenhouse gas because it absorbs heat escaping from the earth and radiates it back like the glass panes in a greenhouse. Consequently, the greater its concentration in the air, the warmer the world climate. The average global temperature today is one degree Fahrenheit (.55°C) warmer that it was one hundred years ago. The seven warmest years on record in more than one hundred years of record keeping have occurred since 1980.

Using the current rate of increase in atmospheric carbon dioxide, computer models predict the global temperature could increase by 4 to 9°F (2–5°C) by the year 2050. This would have disastrous effects on the world. Eventual melting of the polar ice caps would flood much of Florida, Cape Cod, New York City, Los Angeles, and other coastal cities of the world. The rate of global warming would be even greater if it were not for the oceans. It has been calculated that between 50 and 60 percent of the carbon dioxide produced by the burning of fossil fuels is absorbed by the oceans of the world.

The Biological Pump. Phytoplankton continuously take carbon dioxide from the environment and add oxygen in the process of photosynthesis. This has been going on for millions of years and is known as the biological pump. It is a mechanism by which the oceans exert a powerful influence on the carbon dioxide/oxygen balance of the earth. Although most of the carbon dioxide used by phytoplankton is recycled through food webs, some compounds sink to the bottom when organisms die. These may be converted to carbon dioxide by deep sea consumers, then stored for thousands of years until brought to the surface by upwelling deep currents.

The oceans are essential links in the carbon cycle because they hold more carbon than all terrestrial ecosystems combined. According to one estimate, 7 billion tons of carbon enters the atmosphere as carbon dioxide each year. This is mainly from burning fossil fuels. Up to one half of this quantity is absorbed by the oceans, making them a powerful brake on global warming. Conditions that disturb ocean ecosystems, such as pollution, run the risk of disrupting the functioning of the biological pump and accelerating the rate of global warming.

Pollution

Sources. The oceans have long been used as a dumping ground for unwanted waste. When humans numbered only a few hundred million, this may have been a relatively harmless practice. Today, with a growing earth population of over 6 billion, ocean ecosystems are showing definite signs of stress. Pollution is usually greatest along coastlines. One reason for this is that more than half of the people in the world live within 60 miles (100 km) of a coast. Of the ten largest cities of the world, nine are located on sea coasts. Coastal cities have traditionally used the oceans to dispose of garbage and sewage.

Another reason sea water near the coast has a greater degree of pollution is the rivers and streams that empty into the ocean. They carry the pollutants from activities occurring inland—manufacturing, wastes from cities, and agricultural run-off. Included is a vast array of contaminants from toxic industrial waste to raw sewage, and from household detergents to pesticides and farm fertilizers. The distribution of these substances is not restricted to the seashore: ocean currents have allowed some to become worldwide pollutants. Rivers and streams carry 44 percent of all pollutants that enter the oceans.

The prevailing winds also act as transporters of pollutants. A classic example of this is acid rain in the Adirondack Mountains caused by sulfur and nitrogen compounds carried by wind from coal- and oil-burning power plants to the south and west. As a consequence of acid rain, many Adirondack lakes are devoid of life. Pollutants also can be carried into the oceans. Oil is one of the major ocean pollutants, and 10 percent of it comes from atmospheric washout. Heavy metals such as mercury and lead, as well as pesticides and polychlorinated biphenyls (PCBs), are carried by air currents. Land-based operations release these and other substances into the atmosphere, which carries about 33 percent of all pollutants that enter the oceans.

Oil. Small quantities of oil have been seeping into the oceans since they were formed, but most of the contamination has been the result of human activities. Petroleum is the single most important commercial source of energy today, and at least one third of it is shipped by ocean-going oil tankers. Wrecks of these ships, leakage during normal operation, and loading mishaps have been the main sources of pollution. Graphic newspaper accounts of the horrors to wildlife and the despoiling of beaches have followed such well-known accidents as the *Torrey Canyon* in 1967, the *Amoco Cadiz* in 1978, and the *Exxon Valdez* in 1989.

Because oil spills mean loss of revenue to the oil industry, there have been significant improvements in controlling pollution, but oil continues to be a prominent pollutant. As of 1990, 6 million tons of oil were discharged, either deliberately or accidentally, into the oceans each year. It is currently a worldwide problem; petroleum globules have been observed in the Sargasso Sea and in the middle of the Pacific Ocean. As drilling operations on continental shelves increase, it is probable that there will be an increase in contamination.

It is impossible to know with certainty the total effect of long-term oil pollution on ocean ecosystems. Since oil will continue to be an important world commodity in the foreseeable future, the best thing that can be hoped for is better management in its loading, unloading, and transportation. A reduction in leaks and spills is especially urgent in seaports and harbors associated with estuary ecosystems. These important and very sensitive habitats will be discussed in another section.

Heavy Metals: Mercury and Lead. Another product of industrialization is heavy-metal contamination of the oceans. The results of this were illustrated dramatically in 1953 in a small fishing village on Minamata Bay, Japan. At first called Minamata disease, the causative agent was later discovered to be the mercury in wastewater from a polyvinyl chloride factory located on the bay. The mercury was taken up by the shellfish that made up a large part of the village's food supply. More than a hundred people were affected by the disease and almost half of them died. The brain functioning of those who survived was severely impaired. Mercury destroys segments of the central nervous system. Global contamination was indicated in the 1960s and 1970s by the discovery of mercury in the tissues of open sea fish like tuna and swordfish. Although their mercury levels are considerably higher than is natural, they are currently not at toxic levels for humans unless they are eaten in very large quantities.

Contaminants in the atmosphere are normally returned to the land or water in precipitation. In arctic regions they are returned in snow, some of which eventually becomes glacial ice. Analysis of cores from the Greenland ice cap revealed that between 1750 and 1940 there was a 400 percent increase in atmospheric lead. From 1940 to 1967 there was an additional increase of 300 percent. The last increase was mainly from the use of leaded gasoline in automobiles. Lead is a global contaminant that, like mercury, causes brain damage, especially in children. It reaches the ocean mainly by atmospheric washout but also through rivers and streams. Mercury and lead are absorbed from the water and become incorporated, in tiny amounts, into the protoplasm of phytoplankton. They are then passed along in the food web. More research is necessary in order to determine how they influence photosynthesis in phytoplankton, but it is known that mercury is lethal to many species.

Chlorinated hydrocarbons (DDT and PCBs). DDT and PCBs are long-lasting substances that owe their existence entirely to humans. DDT was synthesized first during World War II, and it was hailed as a major breakthrough in the search for a broad-range insecticide. It was used generously and exported to the rest of the world. The first signs of danger from overuse began in the late 1950s with the decline in populations of fish-eating marine birds such as the brown pelican and osprey. The osprey population on Long Island, New York, continued to decline through the 1960s. Populations of peregrine falcons, bald eagles, and other birds also decreased.

PCBs are substances manufactured for use in a variety of products. They have been used as fire retardants in plaster and paint. A major early use was to prevent overheating in electrical transformers, many of which are still in use. Although the manufacture and sale of PCBs in the United States was stopped in 1978, in 2001 the Environmental Protection Agency was still trying to rid the Hudson River in New York State of sediments with high concentrations of these substances. In spite of efforts to contain and dispose of PCB residues, all the oceans of the world are contaminated. In addition to other suspected negative effects on the environment, there is evidence that PCBs are the cause of spontaneous abortions in sea lions and death of shrimp.

The chlorinated hydrocarbons get into the oceans mainly by atmospheric washout. From a very low concentration in sea water, they are incorporated into the protoplasm of phytoplankton. Each zooplankton organism may eat dozens of phytoplankton cells, so the concentration of contaminant grows. A small fish may eat hundreds of zooplankton, again increasing the concentration. Larger fish may eat thousands of small fish, accumulating a still higher concentration. This is called biological or food web amplification, and it is the way that DDT, PCBs, mercury, lead, and other toxic substances reach levels that are harmful to organisms at the top of the food web.

A study on Long Island Sound, near New York City, found the concentration of DDT in the water to be three parts per trillion. In a food chain that went from phytoplankton to zooplankton to small fish to large fish to fish-eating ospreys, the concentration of DDT was increased 10 million times. In high concentrations in osprey and other fish-eating birds, the DDT reduced the amount of calcium carbonate in egg shells. The result was that when the birds sat on, or stepped on the eggs, the shells broke.

The production of DDT in the United States was banned in 1971, and relatively soon thereafter studies showed a thickening in osprey egg shells. Even though it is no longer being used in most of the northern hemisphere, some countries of the world, such as Brazil, are still producing DDT. As indicated by its presence in the tissues of Antarctic penguins and arctic polar bears, it is a global pollutant. Some physicians think its presence in the tissues of most Americans may contribute to the high rate of cancer. The total effect of chlorinated hydrocarbons on phytoplankton is not known, but it is known that DDT is harmful to some species and it reduces the rate of photosynthesis.

The above chemicals are not the only materials that can cause pollution. In addition to known compounds, the chemical industry produces thousands of new substances every year. Marine ecologists have no idea as to the influence that many of these will have on ocean ecosystems.

The Law of the Sea. Mercury and lead are not the only heavy metals that pollute the oceans, nor are DDT and PCBs the only toxic chemical pollutants. There are also plastics. All plastic products are made from small pellets that are transported in bulk by ocean vessels. Spillage during loading and unloading has resulted in worldwide pollution. Most of the public beaches in the world have plastic pellets in the sand. Also plastic products and fragments such as six-pack rings, cargo straps, fishing line, and fishing nets have earned ratings as high as any pollutant for the destruction of marine life. Human treatment of the oceans often appears as though there is a deliberate effort to destroy ocean ecosystems. Deliberate or not, the effect is the same.

Two important questions come to mind: can the oceans be saved, and if so, by whom? These are easy questions to answer but not so easy to accomplish. The species that have already become extinct obviously cannot be restored, but the oceans are far from lost. Experience with Lake Erie, declared dead by some ecologists in 1970, has taught that a body of water can be restored to a degree of health. With 13 million people living along its shores, Lake Erie is still polluted. But by agreement between United States and Canada, many pollutants have been reduced.

The best hope for saving the oceans from further descent into pollution probably lies with the United Nations. As early as 1958, the United Nations held the first in a series of Law of the Sea conferences. By the eleventh conference in 1982, the participants completed a Law of the Sea Treaty. The treaty covers many aspects of ocean management including a provision for each participating country to establish national laws and regulations to prevent and control pollution. Section 11 of the treaty declares the international sea bed and its mineral resources to be the heritage of all people. It further states that any nation with the technology to tap these resources must share the profits with nations not having that technology. In 1998, 130 nations had ratified the treaty. Because of the share-the-wealth section 11, and the opposition of powerful mining forces, the United States has, at this writing, refused to sign the agreement.

The oceans are at least 4 billion years old. They were here long before any life appeared on earth. In the billions of years since life evolved, delicate, well-balanced ecosystems have developed. It is not likely that human activities will ever destroy the oceans, but it is increasingly clear they can destroy ocean ecosystems. This is already occurring to an alarming degree. Many ocean species have become extinct, and others have been so reduced in numbers that they are in danger of becoming extinct. People and governments of the world are beginning to recognize this, but there are still many problems that must be solved. Human and other life-forms depend on the functioning of ocean ecosystems. If the oceans die, all life on earth, as we know it, dies with them.

The Seashore

Where the ocean meets the land there is a constantly changing shoreline. The shore is defined as the area exposed at the lowest tidal level to the highest point washed by storm waves. Kinds of shorelines are highly variable but two common ones are rocky and sandy. The coastal rock layers from Florida to New Jersey are sandstone types that waves have eroded to form the sandy beaches characteristic of this region. Erosion-resistant granite, common in Maine, has produced rocky shores. Glacier-transported material has been deposited by waves to form rocky or sandy beaches in parts of New York and New England south of Maine.

The Intertidal Zone

High tides and storm waves are not uncommon features of the ocean shoreline. These produce an intertidal zone that is a harsh environment for plants, but some species survive there. Along shorelines with rocky bottoms, especially in the northeast, several species of algae, called seaweeds, are commonly observed. These plants must withstand the battering of waves as well as exposure to the drying effect of the air at low tide. Structures called holdfasts anchor them to the rocky bottom and prevent them from being washed ashore. A sticky gelatin-like layer on their outer surfaces enables them to survive exposure to the air. Some seaweeds common along the Atlantic coast, especially in the northeast, are described below.

Sea lettuce (Ulva lactuca, fig. 2.1) is a green alga that grows along the Atlantic coast from Florida to Newfoundland. Although it is normally attached to the rocks in the intertidal zone, it also thrives in the quiet, shallow water of estuaries and tidal pools. It is often fragmented or torn from its attachment and washed ashore. Of all the seaweeds, sea lettuce is probably the one most commonly seen along shorelines and beaches. For more information on this species, see chapter 2.

Rockweed (Fucus vesiculosus, fig. 2.2) is a brown alga characterized by the presence of prominent air bladders that serve as floats when the plant grows in quiet water. It is the most abundant species of rockweed and easy to identify because none of the others have air bladders. It is common in the intertidal zone from North Carolina to Hudson Bay. For more information, see chapter 2.

Broad-leafed kelp or sweet tangle (Laminaria saccharina, fig. 1.4) is a brown alga that may be 1 foot (30 cm) wide and 6 feet (2 m) long. When this plant begins to dry, a sweet white powder accumulates on its surface, thus the name sweet tangle. Its growth form includes a much-branched basal holdfast that attaches it to a rock, a stalk of varying length, and an expanded portion called the blade. The plant is brown because of an accessory pigment called fucoxanthin that masks the green chlorophyll pigment. As it ages, the edges of the blade become ruffled. In autumn the blade becomes leathery. Brown algae are also called kelp. Broadleaf kelp is common from Cape Cod to

1.4. Broad-leafed Kelp *(Laminaria saccharina)*

Newfoundland, Labrador, and Baffin Island. A similar species, *Laminaria agardhii*, is more common south of Cape Cod.

Dulse (*Palmaria palmata*, fig. 2.3) is a red alga that ranges from New Jersey to Newfoundland and Ellesmere Island. Its color may be from deep red to purple to nearly black. Like other seaweeds, it has a holdfast, stalk, and blade. Smaller blades usually develop along the margin of the main one in a palmate manner making it easy to identify. For further information, see chapter 2.

Estuaries

Inland from the upper limit of the shoreline is the coastal region. It is usually defined as the land area that shows signs of influence by the ocean and the marine forces of tides, winds, and waves. Depending on the topography, this zone may extend inland several yards to several miles. Estuaries are common features of the coastal zone of eastern North America. These are shallow, partially enclosed bodies of water freely open to the sea but diluted by freshwater from inflowing streams. The influx of streamwater results in great variation in the salinity of estaurine water. Incoming tides bring ocean water that increases the salinity. It is decreased as the tide ebbs and in times of increased precipitation. Salinity decreases progressively upstream to brackish water at the inland edge of the coastal zone.

Many of the estuaries along the Atlantic coast were formed during the ice ages. Because vast quantities of water were locked up in glacial ice, the level of the ocean dropped several hundred feet. This extended the shoreline seaward and exposed thousands of square miles of land surface. During the hundreds of thousands of years the land was exposed, valleys were eroded by streams flowing to the sea. When the glaciers melted, the level of the ocean rose and the valleys were submerged. Chesapeake Bay, Delaware Bay, and Pamlico Sound in North Carolina are estuaries that originated as drowned valleys. It is in these areas that the ocean ecosystems merge with those of the land.

Estuaries are among the most productive ecosystems on earth. For example, Chesapeake Bay has been an important source of seafood for Americans since colonial times. It is the largest and most productive estuary in eastern North America. It produces more blue crabs than any other area of similar size in the world. It is the richest source of oysters and soft-shell or steamer clams in the United States. Its annual harvest of fish is worth at least $100 million.

The human population in the area surrounding the bay has grown steadily to an estimated 15 million in the year 2000. Unfortunately, the result is that Chesapeake Bay is the most polluted estuary in North America. It is still a productive ecosystem, although a declining one, and many species of plants and animals are threatened by the continuing high rate of pollution.

The bottom sediments of most estuaries consist of mud or sand carried in by streams. Since the water is usually shallow, sunlight can penetrate to the bottom where it sustains a thriving plant community. When viewed from air, these underwater plants give the appearance of submerged grassy meadows. The plants are actually not members of the grass family but have a grasslike form. These submerged plant communities support most of the animal life in the estuary.

Some of the very few flowering plants that can survive in salt water are described below.

Eelgrass (Zostera marina, fig. 1.5) has separate male and female flowers, usually hidden from view, arranged alternately along a portion of the stem, and enclosed in a tubular leaflike structure. It is a perennial with a creeping branching rootstock and a flattened stem. The leaves are long and narrow, rounded to a point at the tip, attached on alternate sides of the stems, and up to 6 feet (2 m) long. Eelgrass is the most widespread and abundant flowering plant growing in saltwater. It can withstand vigorous currents and grows best in deeper water with a bottom of little mud or even pure sand. It is at the base of many estuarine food webs. It is circumboreal in distribution and grows along the Atlantic coast from Florida to Greenland.

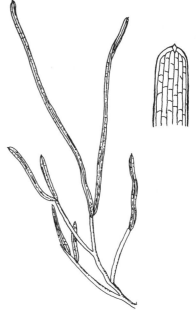

1.5. Eelgrass *(Zostera marina)*

Widgeon grass or ditch grass (Ruppia maritima, fig. 1.6) has tiny flowers with both stamens and pistils in clusters that, at maturity, are raised on long coiled stalks to the surface of the water for pollination. Its leaves are thread-like, attached alternately to a slightly larger, forking stem with a flower cluster at the end of each branch. The stem may be $2^1/_3$ feet (80 cm) long with leaves 2 to 4 inches (3–10 cm) long. The tip of the plant is typically congested with branches. Widgeon grass is a perennial that grows best in quiet waters with muddy bottoms. It grows along both the east and west coasts of North America and inland in areas of salty or brackish water. In estuary food webs, its small leaves are relished by several species of ducks, geese, and swans.

Turtlegrass (Thalassia testudinum, fig. 1.7) is a perennial with male and female flowers on different plants. The leaves form a basal sheath around

1.6. Widgeon grass (*Ruppia maritima*)

1.7. Turtlegrass (*Thalassia testudinum*)

the stem and are shaped like ribbons up to 1 1/3 feet (40 cm) long and 1/2 inch (1 cm) wide. The leaf bases of old leaves persist around the base of the stem giving it a shaggy look. The stems arise from a creeping rootstock. Turtlegrass is a common plant in shallow bays and estuaries along the coast of southern Florida and the Gulf coast. It forms dense underwater meadows, and sometimes during storms great quantities are torn loose and washed up on beaches.

Mangrove Swamps

On the margins of estuaries, marine aquatic ecosystems grade into terrestrial ones. Two ecosystem types are common: salt marshes and mangrove swamps. Salt marshes will be explored in chapter 5. Mangrove swamps are tropical or subtropical wetlands. Mangrove is a general term that refers to an environmental type rather than a particular plant species. There are about eighty unrelated species that make up mangroves throughout the world. They occur in geographic regions where the average temperature does not drop below 68°F (20°C) and the fluctuations in temperature do not exceed 18°F (10°C). The northern edge of mangrove distribution is approximately 29 degrees north latitude.

Mangrove swamps are usually dense stands with many roots and stems that impede water movement. This and a low elevation gradient contributes to the deposition of water-carried sediments. Mangrove swamps develop in bays, lagoons, and estuaries protected from heavy waves and tides that would erode the substrate of mud or sand that supports mangrove seedlings. Mangrove plants evolved more than 100 million years ago from plants in marginal forests. They can grow in freshwater habitats, but they grow best in salt water because there they are free from competition with freshwater plants. The salinity of mangrove swamps varies from lowest where there is an influx of freshwater, to highest along margins where evaporation may concentrate the salt.

In the United States, mangrove vegetation is limited to the intertidal zone between neap and spring tides mostly in south Florida and Puerto Rico. It has been estimated that Florida supports 1,050 square miles (2,730 square km) of mangrove swamps. These form a fringe around south Florida from Cape Canaveral on the Atlantic side to Cedar Key on the Gulf of Mexico side. They are best developed along the southwest coast on the seaward

1.8. Prop roots

margin of the Everglades and Big Cypress Swamp. The mangrove fringe may extend inland for eighteen miles in some areas. It includes the region of the Ten Thousand Islands, some of which have been extensively developed, along that part of coastal Florida. On Marco Island more than a quarter of the mangrove swamps had been eliminated by 1982. These plants are now protected by law from damage or destruction.

Although they are unrelated genetically, many species of mangrove vegetation have developed similar mechanisms for survival. Because ancestral freshwater species had low tolerances for salt, this has been the common factor that has influenced the evolution of mangrove species. They have developed the ability to resist penetration of salt to living tissues and expel that which does enter by salt glands on the leaves. They have developed prop roots (fig. 1.8) that hold the leaves above a fluctuating water level. The prop roots absorb and distribute oxygen to roots that are always submerged in an often oxygen-free substrate. Seeds of mangrove plants germinate while still attached to the parent plant. This keeps them away from salt during germination and allows the root and shoot to develop more quickly when they fall to the ground or are washed ashore. The three main species that make up the mangrove swamps of southern Florida are described below.

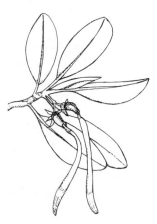

1.9. Red Mangrove
(*Rhizophora mangle*)

Red mangrove (*Rhizophora mangle*, fig. 1.9) grows on the seaward or outer margin of the mangrove swamp and sometimes in isolated clumps in tidal meadows beyond the landward margin. It has small flowers, in clusters of two to several on a long common stalk, that bloom all summer. It has paired, thick, leathery leaves 3 to 6 inches (7–12 cm) long and 1 to 2 inches (2–5 cm) wide. Red mangrove is a small tree, but under good growing conditions it may reach 35 feet (10 m) in height. It has two types of roots: prop roots that arise from the lower part of the stem and drop roots that originate from branches and the upper part of the stem. The characteristic features of red mangrove are the germination of the single seed in each flower and the protruding roots

1.10. Black Mangrove (*Avicennia germinans*)

of the seed that hang from its branches. The roots may be 8 inches (20 cm) or more in length. When the seed drops off the tree, it may embed in the mud or sand where it continues to develop roots and shoots. If it lands in water, it washes ashore and continues growth when it makes contact with solid land.

Black mangrove (Avicennia germinans, fig. 1.10) has greater tolerance for salt than any of the other Florida mangroves. It usually grows on the shoreward margin of the red mangrove zone where the concentration of salt is often higher. It has clusters of small flowers at the tips of stems and branches. Its paired, leathery leaves are 2 to 4 inches (4–9 cm) long and about 1 inch (2.5 cm) wide, and dark green on the upper surface with a whitish undersurface. The upper surfaces of the leaves have salt glands for ridding the plant of excess salt. Black mangrove grows where the ground is usually flooded at high tide. It has prop roots that grow from the stem and underground horizontal roots that produce erect branches called pneumatophores, which absorb oxygen that supports submerged roots growing in oxygen-free substrates.

White mangrove (Laguncularia racemosa, fig. 1.11) has a lower tolerance for salt than any of

1.11. White Mangrove (*Laguncularia racemosa*)

the Florida mangrove species. It grows in a zone inland from the black mangroves where the substrate is flooded only at high tides. It has elongated clusters of small flowers that grow from leaf axils and the tips of stems and branches. The leaves are smaller and thinner than those of black or red mangrove. They are thick and shiny, in pairs, 1 to 3 inches (2–7 cm) long and 1 to 2 inches (2–3 cm) wide. There is a pair of salt glands at the base of each leaf that serves to eliminate excess salt from the plant. White mangrove sometimes grows with black mangrove in inland depressions. The ground beneath them is often covered with pneumatophores.

2

Types of Plants

The word "plant" brings to mind for many people a tree, a house plant, a flower, or a weed. These certainly are all plants, but they are all of the same type. They all produce seeds. The seed plants are the ones we most often see because they are the largest and the most numerous plants on the earth. But there are other types of plants, and it would be difficult to take a walk on the beach, or even in your backyard, without seeing some of them. This chapter explores the different forms of plant life the naturalist is likely to observe at the seashore.

Algae

Algae are classified by the apparently inconsequential quality of color—blue-green, green, red, and brown—which actually indicates a great deal about the internal workings of the cell. They all have relatively simple structures and none of them have roots, stems, leaves, or flowers. The algae have been on earth for more than 3 billion years, and those ancient algae were the ancestors of all modern plants. Along the seashore one can observe all the types of algae mentioned above.

Blue-Green Algae

The blue-green algae are different from the green algae in a number of ways. All algae have chlorophyll pigments, but the blue-greens also contain a bluish-green pigment that gives them their characteristic color. They differ from the green algae also in the arrangement of their genetic material. In green algae, as in most plants and animals, the genetic material or DNA in each cell is enclosed in a central structure called the nucleus. Blue-green algae have no nucleus: instead the strands of DNA are dispersed throughout the cell. In this regard they are more similar to bacteria than to green algae. Newer systems of classification reflect these similarities by classifying bacteria and blue-green algae as cyanobacteria and placing them in a separate group called the Kingdom Monera.

The blue-green algae are abundant in oceanic phytoplankton, but unlike other algae noted above they are all microscopic in size. Most species in this group are blue-green in color, but one is red and its presence in great numbers has given the Red Sea its name. In the field they can be observed as bluish-green or purple patches on rocks in the intertidal zone or on supports of marine wharves.

Green Algae

The green algae are descendants of the blue-greens and are so called because the main pigments that give them their color are the green chlorophyll pigments. They grow in a wide range of both freshwater and saltwater habitats. On the inland margins of the seacoast, freshwater green algae are commonly seen as green slimy masses in roadside ditches, ponds, and streams. When examined under a microscope, these masses appear as delicate green strands, each consisting of many cells attached end to end.

As one approaches the ocean, freshwater forms are replaced by marine species. There is an even greater variety of these than of the freshwater forms. The main constituents of oceanic phytoplankton are green algae called diatoms (fig. 1.3). While these, as well as all freshwater algae, are microscopic in size, some marine forms of green, brown, and red algae are macroscopic.

Along the seashore, a green alga called sea lettuce *(Ulva lactuca,* fig. 2.1) can often be observed. It has a leaflike shape of up to a foot long but is only two cells thick. It is called sea lettuce because it has a superficial resemblance to a leaf of lettuce. The plant usually grows with one end attached to a rock in the zone between low tide and high tide. During storms or in high waves, they are sometimes torn free and washed ashore.

Brown Algae

For naturalists exploring the northern seashores, the most familiar alga observed will undoubtedly be one of the brown algae. The green chlorophyll pigments in these plants are masked by a brown pigment. They usually grow attached to rocks in the intertidal zone in colder sea waters. Brown algae are collectively known as kelp and may be 200 feet or more in length. They are usually attached to rocks by long stalks with leaflike parts containing air bladders to give them buoyancy. Probably the most common seaweed along the Atlantic coast is rockweed *(Fucus vesiculosus,* fig. 2.2), which is characterized by a repeated forking growth with a stalk and an attachment disk. It grows abundantly from North Carolina to Hudson Bay. These plants are often torn free and washed ashore where they look like pieces of dark brown leather.

2.1. Sea Lettuce *(Ulva lactuca)*

One group of brown algae is made up of free-floating forms called gulfweeds or sargassum. They float in the surface waters of an area called the Sargasso Sea, which is roughly the size of the United States, in the southern Atlantic Ocean. It may be the largest area on earth dominated by a single plant genus. One species of gulfweed *(Sargassum fluitans,* fig. 1.2) is abundant in the Gulf of Mexico. During storms, large quantities of this plant are often washed ashore on Gulf beaches and sometimes farther north. Species of gulfweed are easy to identify because they have a stemlike axis with appendages that resemble leaves of seed plants. They also have numerous air bladders that resemble small berries. Although most species of sargassum are tropical, one species *(S. filipendula)*

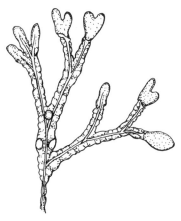

2.2. Rockweed *(Fucus vesiculosus)*

grows as far north as Massachusetts, attached to rocks in the deeper water beyond the low tide level.

Red Algae

The red algae have a red accessory pigment that masks the green of chlorophyll. Their color varies from light pink, to pale red, to reddish purple. The red pigment absorbs light in the short wavelengths of the green-blue range and transfers its energy to the photosynthetic pigments. Short wavelengths penetrate to the greatest depths, so the red algae can grow in deeper water than the green or brown algae. In clear water, they may thrive at a depth of more than 600 feet (180 m). The density of the pigment varies, becoming more dense in deeper water. But even in shallow water, where the pigment is less concentrated, red algae can easily be distinguished from green and brown.

Most red algae are marine; only about 2 percent of the species are adapted to freshwater. They exhibit a great variety of growth forms, some of which are elaborate and beautiful. Although there are several species that are microscopic in size, most red algae are macroscopic. A common species in the intertidal zone from New Jersey to Newfoundland is called dulse (*Palmaria palmata*, fig. 2.3).

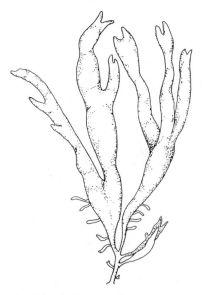

2.3. Dulse (*Palmaria palmata*)

Fungi

In older systems of classification, the fungi were included in the plant kingdom. This may have been because they lack animal characteristics more than that they possess plant features. The fact is they have traits of both plants and animals. For example, they have rigid cell walls like plants, but like animals, they do not possess chlorophyll or make their own food. However, the fungi are a very diverse group of organisms with characteristics that differ enough from both the plant and animal kingdoms to justify placing them in a separate category, the Kingdom Fungi. The ancestors of at least some of the fungi were probably green algae, and, like the algae, they have been on earth 3 billion years or more.

The growth form of the fungi is basically filamentous, consisting of long microscopic threadlike strands. A large number of these strands, called the mycelium, are usually dispersed in the soil or in the dead body of a plant or animal. Most fungi are saprobes, obtaining nourishment by secreting enzymes that digest organic material. At some point in the life cycle of many fungi, the strands of the mycelium grow together in a dense mass that appears aboveground as a macroscopic fruiting body. The function of the fruiting body is to form microscopic reproductive cells called spores. The spores are dispersed primarily by wind, water, or animals, and under adverse conditions they may remain viable for long periods of time. When they fall on a suitable medium, they germinate and grow into a new mycelium.

There are three major groups of fungi that the alert observer will be able to recognize on dunes or in other coastal habitats. These are the slime molds, the sac fungi, and the club fungi.

Sac Fungi (Ascomycetes)

Sac fungi are very important to humans in several ways. On the dark side, they are the causative agents in such diseases as athlete's foot, ringworm, and ergot poisoning, sometimes referred to as St. Anthony's fire. They are also responsible for many serious diseases of crop plants. On the bright side, the yeasts used in baking and in brewing alcoholic beverages are sac fungi. Others give the blue color to some cheeses and the distinctive flavor to Roquefort and Camembert, and they are the source of the antibiotic penicillin.

Some of these fungi have an aboveground fruiting body called an ascocarp. It is frequently shaped like a bowl or a concave disk (fig. 2.4). These range in size from less than $1/2$ inch (1 cm) to more than 5 inches (12.5 cm) in diameter. They can be observed throughout

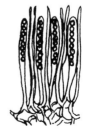

2.4. Sac Fungus

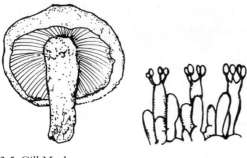

2.5. Gill Mushroom

the summer in moist, shady habitats. The inside lining of the cup or disk may be a dull black or brown, or it may be brightly colored yellow, orange, red, or purple. This colored layer contains asci or sacs in which microscopic spores are produced that will grow into new underground mycelia.

Many ascomycetes are plant or animal parasites in the marine environment. One species has been discovered in the ocean at a depth of over 13,000 feet (3,900 m). Dead marine algae and other plants have been observed to support the growth of these fungi. They may also attack the wood bottoms of boats and wood supports for docks and other saltwater construction.

Club Fungi (Basidiomycetes)

Club fungi are probably more familiar to the general population than any of the other groups of fungi. The name comes from the club shape of the microscopic structures that produce spores, called basidia. Two types of club fungi commonly seen in coastal areas are gill fungi and pore fungi. Gill fungi are common mushrooms that grow in a great range of sizes and colors. A gill mushroom consists of a stalk and an umbrellalike cap (fig. 2.5). On the underside of the cap, thin sheetlike gills radiate out from the central stalk. The spore-bearing basidia are located on each side of the gills.

The color of the spores is an important feature in the identification of mushrooms. Spore color can be determined by making a spore print. This can be done by cutting the stalk near the cap and placing the cap, gill-side down, on white paper, and then covering it with a soup bowl or other convenient cover for a few hours. The spores will collect beneath the gills in lines the color of the spores. To proceed further with identification, refer to the references at the end of this book.

The pore fungi do not have gills. Instead the spore-bearing basidia are located in tiny tubes that open as pores on the underside of the fruiting body. Well-known pore fungi include the bracket or shelf fungi that can be observed on dead trees or logs in coastal woodlands (fig. 2.6). The bracket fungi are important wood-rotting fungi that sometimes attack and kill living trees. Some of them cause dry rot and wood decay in homes and other wood structures. The fungal strands attack by digesting the cellulose in the wood. When the mycelium has become established in the wood, the shelf-like fruiting body develops and airborne spores are dispersed. Some of the bracket fungi are thick and woody with white undersides that are sometimes used as surfaces for artwork.

2.6. Bracket Fungi

A gill mushroom that can be observed on sand dunes from August to October is sandy laccaria *(Laccaria trullisata)*. It has a ruddy cap that develops a central depression as it ages. The gills are reddish purple, becoming a dull red with age.

Rough-stemmed boletus *(Boletus scaber*; fig. 2.7) is another dune mushroom that can be seen among the dune plants from July to November. It has a brownish cap 1 to 5 inches (2.5–12.5 cm) in diameter and a thick stem up to 6 inches (15 cm) long with reddish or brownish scales. Instead of gills on the underside of the cap, this mushroom has tiny tubes. It is an edible mushroom and is sometimes called a tube fungus.

Another group of club fungi that grows in the sand of dunes includes earthstars and puffballs. These do not have the typical stems and caps described for gill mushrooms. In earthstars, a globular fruiting body appears on the surface of the soil. The outer layers peel back in sections forming a star shape. The inner portion has an opening in the top

2.7. Rough-Stemmed Boletus *(Boletus scaber)*

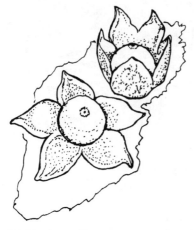

2.8. Earthstar *(Geastrum saccatum)*

through which the mature spores escape. A common earthstar *(Geastrum saccatum*, fig. 2.8) can usually be seen in autumn.

Puffballs are egg shaped or globular when they appear on the surface of the sand. A common puffball *(Lycoperdon perlatum)* is usually seen in the fall and is ordinarily one to a few inches in diameter. Some genera may be as large as 3 feet (90 cm) in length, $2^1/_2$ feet (75 cm) in width and 10 inches (25 cm) high. Puffballs are so called because when they are mature the outer wall breaks and spores are forced out in brown puffs when raindrops or other objects fall on the wall.

Slime Molds

The slime molds are so different that they may not even be related to the other fungi. They are sometimes classified as belonging to a separate kingdom, the Protoctista, which include one-celled protozoans such as amoeba and paramecium. Instead of a filamentous structure, the slime mold is a macroscopic gelatinous mass of protoplasm called a plasmodium, which has numerous embedded nuclei. Like a miniature version of a film monster, it flows very slowly along a surface such as a rotting log, engulfing tiny bits of organic matter that it digests as food. In one phase of its life cycle, it forms stalked, often brightly colored sporangia in which hard-walled spores develop. These are dispersed by wind, and when one falls in a favorable environment it grows into a new plasmodium. Slime molds are most readily observed in moist places where there is an abundance of decaying plant material.

Lichens

Lichens are everywhere: on the bark of trees, seashore rocks, the stones or bricks of buildings, and sometimes even on the backs of seashore tortoises. They are everywhere, that is, except in areas with heavy air pollution. In the centers of heavy industrialization where sulfur dioxide pollution is greatest, there are practically no lichens. This dead zone extends outward for a considerable distance, especially in the direction of prevailing winds. Lichens, then—or their absence—are indicators of air quality.

A lichen is actually not one organism but two. It consists of an alga and a fungus living in very close association. The alga may be a green or a blue-green and the fungus is most often a sac fungus. The algal component manufactures food while the fungus absorbs moisture and mineral nutrients. This kind of association in which there are benefits for both organisms is called mutualism. Most of the body of the lichen is made up of compactly interwoven fungal strands. The alga forms a very thin layer just below the surface of the fungal strands and constitutes about 5 percent of the dry weight. This relationship of alga to fungus is a very complex one that required many millions of years to evolve. Fossil evidence indicates the lichens first appeared on earth in the Mesozoic era, which began about 225 million years ago.

Lichens are able to survive in very harsh environments. They grow on bare rocks where the temperature may reach 122°F (50°C) in summer and in the Antarctic where they survive and may even carry on photosynthesis at -65°F (-50°C). The surfaces on which they grow are usually rather sterile so their chief sources of mineral nutrients seem to be atmospheric dust and rainfall. Since nutrients are limited, the rates of growth are very slow, often less than one millimeter per year. Consequently, a large lichen is probably very old and some may be the oldest living things on earth. A few Arctic lichens have been determined to be 4,500 years old.

Three major growth forms of lichens are easily observed in the coastal zone: crustose, foliose, and fruticose. Crustose lichens are thin and very closely attached or embedded in the underlying surface (fig. 2.9). They can be seen on rocks but cannot be detected by touching. Some of the crustose

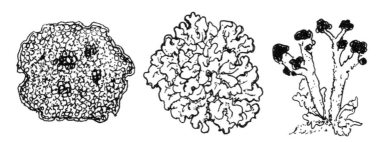

2.9. Crustose Lichen, Foliose Lichen, and Fruticose Lichen

forms are the most tolerant of air pollution. Because they may be present after other forms have disappeared from the area, they are indicator species of polluted air.

Foliose or leaflike lichens are thicker and usually have a central attachment to the substrate with unattached margins (fig. 2.9). During times of drought, the edges tend to roll up tightly and the lichen goes into a state of dormancy. With rainfall or an increase in humidity, it rehydrates and resumes photosynthesis. These and other forms of lichens are usually greenish gray in color, but in alpine conditions and in the Arctic they may be bright yellow, orange, or red. Most foliose and fruticose lichens produce special reproductive structures called soredia that are dispersed by wind. Each soredium consists of a few fungal strands wrapped around one or more algal cells.

Fruticose lichens usually grow attached and perpendicular to the substrate, but some forms hang from tree limbs or other aerial perches. The fruticose forms are the most pollution sensitive of the lichens, and they are the first to disappear as air pollution increases. Familiar examples of these are reindeer lichen *(Cladonia rangiferina)*, British soldiers *(C. cristatella*, fig. 2.9), and old man's beard *(Usnea* spp.). All of these can be observed in some coastal regions protected from salt spray. Reindeer lichens are common in the temperate zone, but they are also an important food source for Arctic animals such as caribou, musk ox, and reindeer. Old man's beard can sometimes be seen on dune shrubs such as beach plum *(Prunus maritima)* and bayberry *(Myrica pensylvanica)*. British soldiers are normally about an inch high with bright red tops.

Some lichens are true marine organisms that grow on rocks in the intertidal zone. They are more abundant along the coast of the North Atlantic than along the northern Pacific in North America. The marine lichens are mostly crustose and often appear as dark brown to black patches on intertidal rocks.

Ferns

The graceful beauty of ferns has always caught the fancy of humans, and almost everyone can recognize a fern. This group of plants is very diverse in size and growth form, and many do not fit the mold of what is thought of as a typical fern. For example, the tiny floating mosquito fern *(Azolla carolini-*

ana, fig. 2.10) may be less than ¹/₄ inch (6 mm) in diameter, and the cloverleaf fern *(Marsilea* spp.) looks like a floating four-leaf clover. At the other end of the size scale, some of the tree ferns in the tropics may be more than 50 feet (15 m) high. Although they have a worldwide distribution in a wide variety of habitats from the equator to the Arctic, 65 to 95 percent of all the fern species grow only in the moist tropics.

2.10. Mosquito Fern *(Azolla caroliniana)*

Ferns and other plants with specialized tubelike cells for conducting substances between the roots and the leaves are called vascular plants. Most of these, including the ferns, have well-developed roots, stems, and leaves. The ferns in the United States and Canada do not have erect aboveground stems but rather have underground ones called rhizomes.

Fern leaves are called fronds. They consist of a stalk or stipe that is attached to the underground stem, with an expanded portion called the blade. In most ferns, the blade is dissected one or more times forming the lacy leaf associated with ferns.

The fern leaf grows in a unique manner. Embryonic leaves that develop on the rhizome are very tightly coiled. As they mature, they unroll from the base in a growth pattern known as circinate vernation. Because the young uncurling leaves look like the neck and head of a violin, they are called fiddleheads. The unrolling of fern fiddleheads is always a welcome and attractive sign of spring. Fiddleheads of several species of ferns are collected in the spring and cooked as green vegetables.

The life cycle of ferns includes two genetically different phases. Haploid spores with half the chromosome number develop in fruit dots or sori on the underside of the leaf. These are dispersed by wind, and when one falls to the ground in a suitable environment, it germinates and grows into a small heart-shaped structure called a prothallus. It is flat, green, and usually ¹/₄ to ¹/₂ inch (6–12 mm) in diameter. The prothallus produces male and female sex cells on its underside. During a warm spring rain or a heavy dew, a sperm cell swims to an egg cell and fertilizes it. The resulting diploid cell grows into a shoot that becomes a frond and a root that gives rise to the underground stem of a new fern.

In earlier times, before their life cycles were understood, ferns were considered to be mysterious plants. Observers assumed they reproduced by seeds, but no seeds could be found. The brown dots on the undersides of the leaves were observed, but no connection was made between these and fern reproduction, so the search continued for the illusive fern seeds. In those times, the unknown was often associated with magic. This may have given rise to the legend that anyone in possession of a fern seed was invisible.

Although none of the ferns are marine or saltwater plants, they are common in woods and fields along the sandy coastal plain. Some can even survive in water that is slightly salty. Netted chain-fern *(Woodwardia areolata)* can grow in semibrackish water and ranges from Nova Scotia to northern Florida. The small floating mosquito fern is common along the coast from Massachusetts to Florida and Louisiana. It grows in quiet waters and sometimes brackish bayous in southern states.

The land surface along the inner margin of the coastal region is often sandy with open vegetation, frequently consisting of pines and oaks. Common ferns in this habitat are bracken fern *(Pteridium aquilinum,* fig. 7.2) and lace-frond grape-fern *(Botrychium dissectum,* fig. 2.11).

Bracken fern is a very robust plant with three-parted, much dissected fronds that usually grows where few other ferns could survive. It produces spores on the undersides of leaflets in fruit dots covered by flaps formed by folded-under margins. It has a rapidly growing underground stem, and it sometimes becomes a weedy nuisance in cultivated fields. Bracken fern is sensitive to temperature and is killed by the first frost, sometimes leaving large patches covered by brown, dead fronds.

Lace-frond grape-fern may grow to 1 foot (30 cm) in height with two branches. One branch near the base of the stem is the frond. It is reddish when it first develops, then turns bronze in autumn. This persists through the winter. The frond has three parts, is much dissected, and up to 3 inches (8 cm) long. The other stalk of the plant has a branching cluster of spore-bearing structures

2.11. Lace-Frond Grape-Fern *(Botrychium dissectum)*

at its tip. This fern grows in a wide variety of habitats, but it may be seen along the coastal plain from Nova Scotia to Florida.

Gymnosperms (Conifers)

The gymnosperms are vascular plants that bear seeds in cones. The term gymnosperm means naked seed and refers to the fact that gymnosperm seeds are not enclosed in a fruit as are those of flowering plants. The gymnosperms have been on earth for about 325 million years. Their ancestors were vascular plants that evolved from green algae. They reached their peak of development during the time of the dinosaurs when they were the dominant vegetation on the earth. Since that time they have declined as flowering plants have become the dominant vegetation.

2.12. Bald Cypress *(Taxodium distichum)*

Conifers

The most numerous and best-known gymnosperms are the conifers. These include such familiar coastal plain trees as pitch pine *(Pinus rigida,* fig. 4.20), pond pine *(P. serotina,* fig. 3.10), bald cypress *(Taxodium distichum,* fig. 2.12), and Atlantic white cedar *(Chamaecyparis thyoides,* fig. 2.13). Although most conifers are evergreen, bald cypress sheds its leaves in autumn. A few conifers are shrubby but none are herbaceous.

The leaves of the conifers are typically needle shaped, as in the pines, or they may appear to be overlapping scales, as in Atlantic

2.13. Atlantic White Cedar *(Chamaecyparis thyoides)*

white cedar. In pines most of the needles are attached on short spur branches in clusters of two to five. Although a new crop of leaves is produced each spring in conifers, all of last year's leaves are not shed before winter. Leaves remain on the tree from one to five years and are shed gradually throughout the year. The shape, type of attachment, length, and texture of the needles are all important features in identifying species of conifers.

Like ferns, the life cycle of conifers can be described as the alternation of haploid and diploid phases. Two type of cones are produced, male or pollen cones and female or seed cones. Haploid male and female sex cells develop in the cones.

The pollen grain is transported by wind to the female cone where it develops a pollen tube that carries the sperm to the egg cell. The sperm and egg unite to form a diploid cell that grows into an embryonic tree. The embryo is surrounded by a layer of nutritive tissue and a hard outer covering. This is the seed that may become one of next year's crop of conifers. In most conifers, pollination occurs in May and the seeds are shed in the autumn. In pines they are shed in autumn the year after pollination. Each scale of the female cone bears two seeds with flat winglike projections that aid in dispersal by wind.

Flowering Plants

The flowering plants are referred to as angiosperms, which means "seeds in a receptacle." This alludes to the fact that their seeds develop inside the ovary, which matures to become the fruit. The flowering plants are the most recently evolved of all the plant groups. Their oldest fossils are about 130 million years old, and their ancestors were ancient gymnosperms that are today extinct. The time when dinosaurs roamed the earth was the age of gymnosperms, but today we are in the age of angiosperms. They dominate world vegetation with the greatest number of individuals and the greatest number of species. Unlike the gymnosperms, they grow in a variety of forms including trees, shrubs, vines, herbs, and nongreen parasites.

Angiosperms are true land dwellers as are the mammals of the animal kingdom. In ferns, water is required for the sperm to swim to the egg in order to complete the life cycle. They resemble the amphibians of the animal world in this regard. In angiosperms the link with aquatic ancestors has been severed. The pollen with sperm cells is delivered to the egg cell by wind, insects, or some other animal pollinator; no water is necessary.

Mammals, birds, and insects have evolved in close association with flowering plants. The rise of herbivorous mammals was dependent on the development of grasses and other herbs. Birds have evolved with their main sources of food, the seeds and fruits of flowering plants or the insects that feed on leaves and fruits. Insects and angiosperms have a remarkable history of codependence and coevolution. The domestication and subsequent cultivation of certain flowering plants was the initial step in the development of modern civilization.

The flower is the organ of reproduction for the angiosperms (fig. 2.14). A typical flower consists of an outer ring of green leaflike parts called sepals. Collectively the sepals make up the calyx. Its function in the bud is the protection of the delicate inner parts. Inside the calyx is the corolla, which is made up of individual parts called petals. The corolla in many flowers is brightly colored and associated with sweet-smelling nectar glands, features that attract insects and other pollinators. Inside the corolla is a ring of stamens, the male parts of the flower, which consist of a slender stalk, the filament, topped by an anther. In the center of the flower is the female reproductive structure, the pistil (there may be more than one), which is made up of an enlarged lower part, the ovary, an elongated neck, the style, topped by a sticky surface, the stigma, which receives the pollen grains. One or more ovules develop inside the ovary, and these will eventually become seeds.

As in other plant groups, angiosperms alternate diploid and haploid phases as part of their life cycles. The anther produces haploid pollen grains with half the chromosome number, which carry the male gametes.

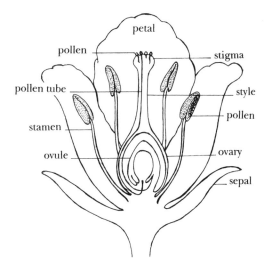

2.14. Angiosperm flower

Each of the ovules inside the ovary has a haploid egg cell. During pollination, the pollen grain is transported to the stigma where it germinates. A pollen tube containing a sperm cell grows through the style and fertilizes the egg cell. The resulting diploid cell develops into an embryonic plant. Food-storage tissue with a hard outer coat forms around the embryo. At this point the ovule becomes a seed.

Monocots and Dicots

There are two basic groups of flowering plants, the monocotyledons (monocots) and the dicotyledons (dicots), that are fairly easy to distinguish in the field. Since there are many more dicots than monocots, the plants most often seen are dicots. Cotyledons are seed leaves and as the names suggest, monocots have one and dicots have two. A seed usually consists of a seed coat, food-storage tissue, and the embryonic plant with its first leaves. Sometimes all the stored food is converted into the cotyledons as in peanuts, beans, and peas. The time to observe cotyledons is shortly after the seed germinates; the first structures to appear above the ground are the cotyledons, and whether there is one or two will be obvious.

2.15. Perennial Salt-Marsh Aster (*Aster tenuifolius*)

There are features other than seed leaves that can be used in the field to distinguish between these two groups. The leaves of monocots have veins that are parallel from the base to the tip of the leaf; in dicot leaves, the veins are branched into a network. The most reliable and easiest way to distinguish between these two groups is by the number of flower parts. In monocots, the flower parts are in multiples of three. A typical species could have three sepals, three petals, and six stamens.

The flower parts of dicotyledons are in multiples of four or five. Almost all of the trees and shrubs and most of the herbs are dicots. A common flower type of these plants could have five sepals, five petals, and five to ten stamens.

The aster family is the largest family of flowering plants and the most highly evolved of the dicotyledons. The flowers of this family are very small and so tightly clustered that each cluster gives the appearance of a single flower. For example, perennial salt-marsh aster *(Aster tenuifolius,* fig. 2.15) looks like a flower with purple petals. Actually each of the "petals" is a flower and the center is made up of many tiny flowers each consisting of a five-lobed corolla, five stamens, and a pistil with a curling two-part stigma.

The number, shape, color, and location of flower parts are the chief traits used in the identification of herbaceous plants. Leaf characteristics are the most important features used in the identification of trees and shrubs. These will be discussed further in chapter 8.

3

Adaptations for Survival

Genetic Variability

Asexual Reproduction

Most species of plants produce offspring by both asexual and sexual means. In asexual reproduction, there is no union of male and female sex cells. Consequently the genetic makeup of the offspring is exactly the same as that of the parent. It can be as simple as a piece of the parent plant breaking off and growing into a new plant. This type of fragmentation is common in the seaweed sargassum, or gulfweed *(Sargassum fluitans,* fig 1.2). Another simple form of asexual regeneration is exhibited when new shoots sprout from creeping underwater stems of eelgrass *(Zostera marina,* fig. 1.5) or narrow-leaved cattail.

In the most complex form of asexual reproduction, seeds are produced without the union of male and female sex cells. Seeds are normally the result of sexual reproduction, but in some species, such as prickly pear *(Opuntia humifusa,* fig. 3.1), a diploid cell in the ovary grows into an embryo that will eventually become a seed. This type of nonsexual reproduction has the advantage of seed-dispersal mechanisms for spreading into new areas. The seeds will germinate and, like other forms of vegetative reproduction, grow into a plant that is an exact genetic duplicate of the parent plant.

Plant species that reproduce only by asexual means are evolutionary dead ends. In the absence of sexual reproduction, they cannot present a variety of genetic combinations for selection by a changing environment. Because they have only one combination, the only response they can make to climatic change is extinction.

Sexual Reproduction

In sexually reproducing species, there is a reshuffling of genes with each generation, providing a variety of genetic types. As a result, when there is a substantial change in the environment, although some members of the species will die, others may have genetic traits that are advantageous for survival in the changed conditions. It is this quality that led to the origin of sexual reproduction and to its persistence in most living things today. All groups of plants—including the algae, mosses and liverworts, ferns, coniferous plants, and flowering plants—have well-developed sexual systems. Since the flowering plants are the ones most commonly seen along the seashore, their sexual reproduction will be discussed in more detail.

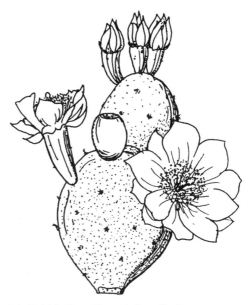

3.1. Prickly Pear *(Opuntia humifusa)*

Methods of Cross-Pollination

Plant populations with the greatest genetic variability are best adapted for long-term survival. In flowering plants, the key to maintaining the greatest amount of variability is cross-pollination. Pollination is the transfer of pollen, which carries the male sex cells to the pistil, where the female sex cells are located. Most flowering plants have both male and female parts in the same flower. Cross-pollination occurs when the pollen from one plant is transferred to the pistil of another. When pollination occurs on the same plant, there is less variability in the offspring than when the pollen is from another plant. Numerous growth habits have evolved that promote cross-pollination.

Animal Pollinators

Insects

The most important agents of cross-pollination are insects. They have been associated with flowering plants for at least 60 million years. During this time, many specialized relationships have evolved in which plants and insects are mutually dependent on one another. Most of the time, cross-pollination by insects is accidental. Visits to flowers are usually for nectar or pollen or both as sources of food. Insect-pollinated plants have sticky pollen that adheres to the body of the pollinator. When the insect visits another flower of the same species, it accidentally brushes against the stigma and the pollen is transferred.

Bee pollination. Bees are the most important of the insect pollinators. There are at least twenty thousand species, all of which must visit flowers for food. Plant species pollinated by bees have evolved special types of flowers that are easy for bees to find and on which they can land. These insects cannot recognize the color red but they can see ultraviolet light, which is invisible to the human eye. Flowers that have developed in response to bee pollinators are usually yellow or blue. Bees and flies are attracted to the bright yellow clusters of seaside-goldenrod *(Solidago sempervirens,* fig. 4.11).

Moth and butterfly pollination. Moths and butterflies are very important pollinators. Most moths are night-flying creatures that have coevolved (developed in response to one another) with night-blooming plants. Since bright colors are not visible at night, most moth flowers are white or of a pale color that will stand out against a dark background. Moths have a well-developed sense of smell and flowers that attract them emit powerful fragrances only after sunset. Both moths and butterflies have long sucking tongues that permit them to reach the nectar in narrow tubular flowers. Included among moth-pollinated flowers is seaside evening-primrose *(Oenothera humifusa,* fig. 4.16).

Flowers pollinated by butterflies are usually showy, fragrant, and day-blooming as are flowers pollinated by bees. Unlike bees, some butterflies can see the color red and visit red and orange flowers as well as blue and yellow ones. Lance-leaved milkweed *(Asclepias lanceolata,* fig. 3.2) grows in brackish salt marshes along the Atlantic coastal plain. It has showy yellow to orange-red flowers, and it is often visited by butterflies. One group of moths, the hawkmoths, are active during the day and may visit the same flowers as butterflies.

Fly and beetle pollination. The food of beetles and flies is frequently decaying fruit, dung, and dead animals. Plants that have been influenced by these insects in their evolution often have flowers with the unpleasant odors of rotting tissue. The sense of smell in beetles is more highly developed than the sense of sight, and the flowers they visit are usually not brightly colored. Most beetles do not have mouth parts suited for obtaining nectar, especially from tubular flowers, so they feed on flower parts or pollen. There are at least thirty thousand species of plants pollinated by beetles with more being discovered each year. Beetles are frequently observed on the flowers of Japanese rose *(Rosa rugosa,* fig. 3.3). This rose grows in other habitats, but it is common on the sand dunes of Cape Cod where it is called saltspray rose.

3.2. Lance-Leaved Milkweed *(Asclepias lanceolata)*

Birds

In different parts of the world, many species of birds are specialized to feed on flower parts, flower-eating insects, or nectar. As with most insect pollinators, cross-

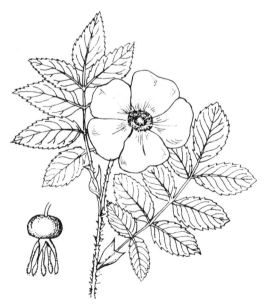

3.3. Japanese Rose *(Rosa rugosa)*

pollination by birds is accidental. In North America, the most important bird pollinators are the hummingbirds whose most important source of food is nectar. They have long slender beaks that can reach the base of the longest tubular flowers. Hummingbirds have a well-developed sense of color and can see reds but have a very poor sense of smell. Consequently, plants pollinated by hummingbirds often have red flowers with little or no odor.

Hummingbirds sometimes pollinate swamp rose-mallow *(Hibiscus moscheutos,* fig. 3.4) in salt—or brackish-water swamps along the Atlantic coast.

Genetic Safeguards

Even though cross-pollination by insects or birds is usually very dependable, there is still the possibility that pollen from the anther could reach the stigma on the same flower. This is called self-pollination, and many plant species have physiological or structural features to keep it from happening. One way has been through the development of different genetic strains within a species. The pollen of one strain is physiologically rejected by the stigma of any flower on the same plant. It must reach the stigma of another plant before it will grow a pollen tube that will result in the production of a viable seed. This is called self-incompatibility and is common among species of wild plants.

A strategy to avoid self-pollination in some insect-pollinated plants is the growth of pistils and stamens with different lengths. Some species have two types of flowers in approximately equal numbers: those with long-styled pistils and short stamens and those with short-styled pistils and long stamens (fig. 3.5). This greatly reduces the probability of pollen reaching the stigma in the same flower. However, as added insurance, this feature is usually accompanied by self-incompatibility.

3.4. Swamp Rose-Mallow *(Hibiscus moscheutos)*

ADAPTATIONS FOR SURVIVAL / 49

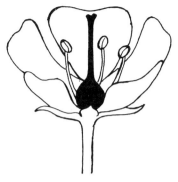 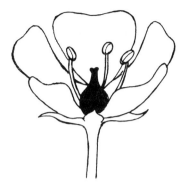

3.5. Type 1 flower; type 2 flower

Fertile seeds can be produced only when pollen from a type 1 flower reaches the stigma of a type 2 flower.

Separation of the Sexes

Some plants have the stamens (male parts) and pistils (female parts) on separate flowers. These are called monoecious plants. Sometimes the male and female flowers are in different locations on the same plant, as in narrow-leaved cattail *(Typha angustifolia,* fig. 3.6). In this plant the male, or staminate flowers, are in an elongated cluster at the top of the stem separated by a small gap from the female flowers in an elongated cluster beneath them. Other species that have separate male and female flowers on the same plant are eelgrass, turtle grass *(Thalassia testudinum,* fig. 1.7), and spearscale *(Atriplex patula,* fig. 5.11). Although self-pollination is less likely in these species, it could occur unless they are self-incompatible.

In other species the individual plants bear either male or female flowers, but not both. These are called dioecious plants, and this arrangement is a guarantee that self-pollination can never occur. The disadvantage is that only half of the population can produce seeds since male and female plants are present in about equal numbers. Species with unisexual flowers

3.6. Narrow-Leaved Cattail *(Typha angustifolia)*

include spike-grass *(Distichlis spicata,* fig. 5.3), marsh elder *(Iva frutescens,* fig. 5.4), and bayberry *(Myrica pensylvanica,* fig. 4.18). Plants with unisexual flowers, both monoecious and dioecious, are more common among the wind-pollinated species.

Wind Pollination

Wind is an extremely inefficient agent of pollination. Whether or not an individual pollen grain reaches a stigma is purely a matter of chance. So in order to ensure fertilization, wind-pollinated plants produce great quantities of pollen. So much is produced that even the most remote stigma is likely to be dusted. Only in this way can a seed crop big enough to sustain the species be assured.

Since their evolution was not influenced by animal pollinators, wind-pollinated flowers do not have colorful petals, do not produce nectar, and do not have a fragrance. Some wind-pollinated seashore plants are coast-blite *(Chenopodium rubrum,* fig. 5.10), southern sea-blite *(Suaeda linearis,* fig. 5.7), and beach wormwood *(Artemisia stelleriana,* fig. 4.8). Usually these species do not produce enough pollen to be serious causes of allergy reactions. The stigmas of wind-pollinated flowers are usually extensively branched, exposing the maximum amount of surface. This increases their efficiency in trapping pollen grains. The ovary usually has only one ovule, so only one pollen grain is necessary for successful pollination.

Seed Dispersal

A discussion of seed dispersal cannot be complete without a clear understanding of the relationship between fruits and seeds. Seeds develop within the ovaries of flowering plants, and the ripened ovary is a fruit. Ovaries can be divided into two kinds, fleshy and dry. Fleshy fruits have thick walls that are sometimes colorful when the seeds are mature. Beach-plum *(Prunus maritima,* fig. 3.7) has a blue to reddish-purple fleshy fruit. Dry fruits are those in which the ovary wall is usually dry when the seeds are mature. Beach pea *(Lathyrus maritimus,* fig. 4.7) and lance-leaved milkweed are plants with dry fruits. In some dry fruits the ovary contains only one seed, and at maturity the ovary wall becomes part of the seed coat. Members of the aster family all have seeds that botanists classify as fruits.

Most seeds fail to complete their evolutionary mission, the growth of a new plant. Consider, for instance, the attractive sand dune species seaside-goldenrod. Each plant normally has sixty to a hundred or more flower heads, each of which produces a minimum of thirty seeds. If all the seeds of every flower head grew to a mature seed-producing plant, the resulting number of seaside-goldenrods would soon be more than could be supported by the entire Atlantic coastal region.

3.7. Beach-Plum (*Prunus maritima*)

In spite of this high rate of failure, seeds continue to be the most important means of reproduction and dispersal among flowering plants. They provide the plant with mobility, allowing the species to colonize new areas and increase its range of distribution. Still another advantage of seed dispersal is that seedlings escape from competition with the parent plant. In response to these advantages—and perhaps others—a variety of seed-dispersal mechanisms have arisen. The two most common agents of dispersal are wind and animals.

Dispersal by Animals

Animals are the most effective agents of seed dispersal. There are at least two reasons for this. First, migrating birds and mammals move at predictable seasonally regulated intervals. Over a long period of time, this could result in the coevolution of plants with seeds that are mature during migration. Second, since mammals usually move from one favorable environment to another, the seeds they transport are likely to be deposited in another area that is favorable for their germination and growth. Dispersal is by three methods: (1) ingestion, (2) adherence to the outer surface of fur, feathers, or feet, and (3) transportation and storage as a food reserve.

Ingestion. Fleshy fruits have evolved chiefly as organs for seed dispersal. Animals are attracted to them as food sources, and seeds are transported in the intestines of the animals. Many seeds pass through animal digestive tracts unharmed. By one estimation, fruit eaters are responsible for seed

3.8. Cocklebur *(Xanthium strumarium)*

dispersal in about 12.5 percent of the flowering plants in northeastern North America. Birds are the most important of these, but mammals and reptiles are also fruit eaters. In over 70 percent of the plants that have bird-disseminated seeds, fruit ripening coincides with the onset of fall bird migration. A seashore species that has seeds dispersed by birds is bayberry.

Adhesion. Seeds dispersed in this manner are usually from herbaceous plants and have hooks, spines, or sticky surfaces. A sand dune plant with seeds in a spiny bur is dune sandspur *(Cenchrus tribuloides,* fig. 4.12). Another bur with hooked spines frequently seen growing on sea beaches is cocklebur *(Xanthium strumarium,* fig. 3.8). The seeds of these species readily become attached for a free ride on the fur of any passing animal or the clothing of any field naturalist.

Many salt-marsh plants produce small seeds with no apparent special structures for dispersal. The wide distribution of some of these species is the result of dispersal by water birds. Their adaptation is their size: small seeds will stick to the mud on the feet of wading birds. The wide distribution along the Atlantic coast of salt-marsh agalinis *(Agalinis maritima,* fig. 5.9) may be the result of this type of dispersion. Dispersal by this method was recognized by Charles Darwin, who was probably the first to make a count of the seeds in the mud on the feet of wild ducks.

Dispersal by Wind

Wind is the second most important agent of seed dispersal. It is less efficient than animals for two reasons: it is highly variable and unpredictable, and it may not be present at the best time for dispersal; and dispersal is random, so many seeds fall in areas unsuitable for germination. Despite these shortcomings, species with seed modifications for wind dispersal are common. A few of these modifications are described in the following paragraphs.

Size. Very small seeds can be seen as an adaptation for dispersal by wind. The smaller the size, the greater the ease of dispersal. The seeds of tall wormwood *(Artemisia campestris,* fig. 3.9) are small enough to become airborne in a moderate breeze.

Parachutes. A common adaptation for dispersal by wind is a tuft of hairs that functions as a parachute. This dispersal method is found in several herbaceous seashore plants, including narrow-leaved cattail, perennial salt-marsh aster, and seaside-goldenrod. Some species have seedpods that open to release seeds with parachutes of hair, such as lance-leaved milkweed. About 16 percent of American plants are dispersed by seeds with parachutes.

Wings. Thin membranous wings are effective mechanisms for dispersal by wind. These slow the rate at which the seed falls, giving the wind time to carry it—sometimes for long distances. Pond pine *(Pinus serotina,* fig. 3.10) and Atlantic white cedar *(Chamaecyparis thyoides,* fig. 2.13) are seashore trees with winged seeds. They whirl like tiny propellers and in strong winds may be carried for several miles.

Tumbleweeds. In some species, the whole plant serves as a seed spreader. These plants have a highly branched, bushy growth habit that gives them a globular shape. When the seeds are mature, the plant breaks at the base and rolls with the wind, scattering seeds as it goes. A seashore plant of this type is saltwort *(Salsola kali,* fig. 6.5).

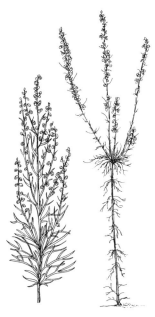

3.9. Tall Wormwood *(Artemisia campestris)*

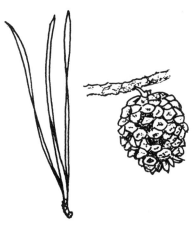

3.10. Pond Pine *(Pinus serotina)*

4

Sand Dunes and Beaches

The Atlantic Coastal Plain

The Atlantic coastal plain is a level strand that extends from Cape Cod to Florida and continues as the Gulf coastal plain to Texas and Mexico. It varies in width, being narrowest in the northeast and becoming much broader southward. The rolling hills of the Piedmont region mark its inland margin. The coastal plain is an extension of the continental shelf, and over the past several million years it has been above and beneath the waves more than once. The sandy bottom of the last submergence constitutes the substrate of the coastal plain today. The shoreline consists of almost continuous sandy strands with recreational beaches along some stretches that are world famous.

These sandy beaches provide distinctive habitats for the growth of plant communities. They comprise a narrow maritime band that seldom extends more than a mile (1.6 km) from the ocean. It is an environment where wind, shifting sand, and salt spray are the controlling factors. These features are so prevailing that some species of plants grow from north to south throughout this coastal fringe. The environmental conditions here are not unlike those of temperate-zone beaches the world over. As a result, many of the plant types that grow on North American beaches have a global distribution. In the following sections, the environmental factors and the plant communities of sandy beaches and dunes of the Atlantic and Gulf coasts are explored.

A Dune Is Born

When wind blows across a sandy expanse, it picks up grains of sand and bounces them along the surface, dislodging other grains that bounce along and dislodge still others. Any object in the path of the wind will reduce the wind's speed, and sand will be dropped around the object. It may be a rock, a log, a single plant, or a clump of plants. A small pile of sand is formed and an embryonic dune is born. As this small mound grows, it merges with other mounds, which merge with still others until eventually a hillock is formed.

4.1. Beach-Grass (*Ammophila breviligulata*)

It has a ridge with a gentle sloping windward side and a very steep leeward side. Once it is initiated, the dune itself becomes an obstacle to the wind, causing the deposition of more sand and the continued growth of the dune.

A dune will continue to grow as long as the wind can transport sand to the top. The sand grains are rolled up the windward slope and tumble down the leeward side, causing the dune to move with the wind. The movement is usually only a few feet per year, but it is relentless, and if the sand is not stabilized it will bury everything in its path. In coastal environments dunes will continue to move until they are anchored by a plant cover that binds the sand. By the time this happens they have moved inland, and smaller frontal dunes have formed on the seaward side.

Along the Great Lakes, Lake Champlain, and the northeastern Atlantic coast as far south as Cape Hatteras, North Carolina, the plants most effective for dune stabilization are strains of beach-grass (*Ammophila breviligulata*, fig. 4.1). On the Pacific coast a related species, maram grass (*A. arenaria*), is an effective dune stabilizer. When beach-grass has become established, the dune can continue to grow if there is a source of sand. The grass reduces the velocity of the wind so that sand grains are deposited and no longer roll over the leeward side. The dune is stabilized and no longer moves with the wind. South of Cape Hatteras, North Carolina, the main dune stabilizer is sea-oats (*Uniola paniculata*, fig. 4.2). Beach-grass may be

56 / A NATURALIST'S GUIDE TO SEASHORE PLANTS

up to 3 feet (1 m) tall with flowers in very dense, more or less cylindrical clusters about an inch (2.5 cm) in diameter. Sea-oats are usually 3 to 8 feet (1–2.5 m) tall with more open flower clusters.

The Dune Environment

A sand dune is a harsh environment for plant growth. One of the reasons is that sand is very poor in mineral nutrients. Most of the minerals that plants must have to grow are held on the surfaces of microscopic soil particles called micelles, which are most common in finely textured soils that contain clay, silt, and decomposed organic matter. Sand has a very coarse texture with little decomposing organic matter and few mineral-holding micelles. Water droplets carried from breaking waves and rainfall are the main sources of mineral nutrients for plants growing on coastal dunes.

4.2. Sea-Oats *(Uniola paniculata)*

Obtaining enough water to survive is another problem of plants growing on sand dunes. The water in soil that is used by plants is held as a film around the soil particles. The thickness of the film is about the same regardless of the size of the particle. The coarsely textured sand cannot hold as much water as soils made up of more numerous smaller particles. To make matters worse, on sunny days in summer the surface temperature on a sand dune may get to 115°F (46°C) or higher. Plants that grow on dunes, such as beach-grass, sea-oats, sea-rocket *(Cakile* spp.), and seaside spurge *(Euphorbia polygonifolia,* fig. 4.3), have root systems that penetrate to a depth below the hot, dry surface sand. The main source of water for these plants is rainfall that percolates downward to the root level. During long dry spells, the condensation of dew on sand below the surface of the dune may contribute to the survival of dune species.

Constantly shifting sand presents another challenge to plants growing on sand dunes. If all other factors were favorable, periodic burial by sand would be enough to keep most plants from growing there. However, being buried by sand actually stimulates beach-grass and sea-oats to more vigorous growth. These plants have horizontal stems under the sand that send up branches as they grow. As the wind blows over the dune and sand builds up around the upright green branches, they grow longer, always keeping their leaves above the sand.

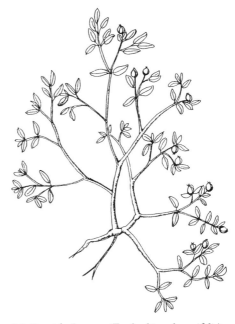

4.3. Seaside Spurge *(Euphorbia polygonifolia)*

On northeastern dunes, upright branches from the underground stems of beach-grass sometimes resemble planted rows of plants. If the leaves are buried, buds in the leaf axils begin to grow and emerge from the sand as new shoots. Investigations have shown that the growth of beach-grass can keep pace with an accumulation rate of as much as 3 feet (1 m) of sand per year. In both beach-grass and sea-oats, the continued growth of stems and leaves are accompanied by the formation of extensive root systems that bind the sand. These plants are well qualified to survive in sand dune environments. The genus name for beach-grass, *Ammophila*, appropriately means "lover of sand."

Plants of the Dunes

Beach and Frontal Dunes

In addition to the hazards described earlier, plants that inhabit beaches must also have a tolerance for salt spray. Although it is the main source of mineral nutrients for coastal dunes, salt spray creates a very hostile environ-

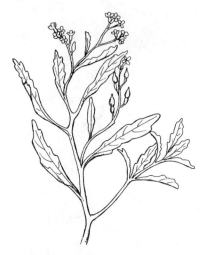

4.4. Sea-Rocket *(Cakile edentula)*

ment for most plants. The chief source of salt spray is bursting bubbles from the wash as waves break on the beach. Fine droplets of water become airborne, and then evaporation concentrates the dissolved salts. The most abundant salt in sea water is sodium chloride (table salt). It is the chloride that is toxic to plants. Only those with a high degree of resistance to salt spray grow on the beach and especially the windward sides of the dunes.

Below the wave line, the beach is constantly being restructured by erosion and deposition so that plant growth is not possible. Between the wave line and the windward face of frontal dunes, a few hardy species can withstand the sometimes intense salt spray. Among these are two species of sea-rocket. A species with northern distribution *(Cakile edentula,* fig. 4.4) grows from Labrador to North Carolina, and a southern one *(C. harperi)* is found on beaches from South Carolina to Florida and Louisiana. These species have similar erect growth forms, but there are other species that range from the Florida Keys to Louisiana with a sprawling growth.

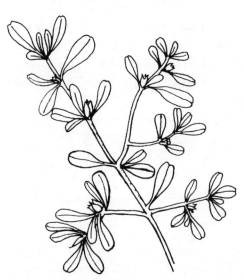

4.5. Sea-Purslane *(Sesuvium maritimum)*

Sea-purslane *(Sesuvium maritimum,* fig. 4.5) can be seen on beaches from Long Island, New York, to Florida and Texas. It has numerous branches up to 16 inches (40 cm) long and pink flowers about $1/_4$ inch (6 mm) across, with five stamens. Another species of sea-purslane *(S.*

portulacastrum) grows on beaches from North Carolina to Florida. It is recognizable by its few branches, some of which may be 6 feet (2 m) long, and flowers with numerous stamens.

A semishrub that sometimes grows on the beach from Virginia to Florida is sea-elder *(Iva imbricata,* fig. 4.6). It is a member of the aster family and has cream-colored flower heads. Sea-rocket, sea-purslane, and sea-elder all have thick leaves and stems that hug the sand. Sea-elder has a perennial rootstock that produces new shoots each year, but sea-rocket and sea-purslane are annuals that grow from seeds every spring. Sea-rocket has flowers with four white to pale purple petals that appear from early summer to September. Sea-elder and sea-purslane bloom from late summer to autumn. In more southerly locations, flowering is likely to begin earlier and continue later.

Plants growing on the windward sides of frontal dunes are exposed to the full force of the wind and salt spray. The growth of beach-grass and sea-oats seem to be unaffected by these factors. Plants of the upper beach may also grow with the dune grasses. Other species sometimes observed on the frontal dunes north of North Carolina are seaside spurge, beach pea *(Lathyrus maritimus,* fig. 4.7), beach wormwood or dusty miller *(Artemisia stelleriana,* fig. 4.8), false heather *(Hudsonia tomentosa,* fig. 4.9), and sea-beach sandwort *(Honckenya peploides,* fig. 4.10). Seaside spurge, beach pea, sea-rocket, and beach-grass are also common on the beaches and dunes around the Great Lakes. Brief descriptions of these plants are given below.

Seaside spurge is a small, sprawling plant. It has red stems, pale green leaves, and milky juice.

Beach pea has purple flowers, com-

4.6. Sea-Elder *(Iva imbricata)*

4.7. Beach Pea *(Lathyrus maritimus)*

4.8. Beach Wormwood
(*Artemisia stelleriana*)

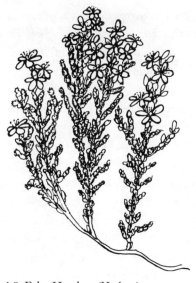

4.9. False Heather (*Hudsonia tomentosa*)

4.10. Sea-Beach Sandwort (*Honckenya peploides*)

4.11. Seaside-Goldenrod
(*Solidago sempervirens*)

pound leaves with a tendril at the tip, and a stem that creeps over the sand. The pealike seedpods contain edible peas when they are very young. Beach pea also grows on beaches of the Pacific coast.

Dusty miller has numerous flower heads and pale green, highly dissected leaves covered with white, woolly hairs.

False heather is a shrubby evergreen that grows in low spreading clumps. It has small, overlapping, densely hairy leaves that cover the stems.

Sea-beach sandwort has small white flowers and fleshy stems with bright yellow-green, oval, pointed leaves. It grows in low clumps that spread over the ground like a carpet. Sea-beach sandwort also grows along the Pacific coast from Alaska to California.

4.12. Dune Sandspur *(Cenchrus tribuloides)*

Leeward Sides and Swales

In areas where there are frontal dunes and back dunes, the windward sides of the back dunes receive almost as much salt spray—and so have the same vegetation—as the windward sides of the frontal dunes. This is especially true when the back dunes are high. The environment is less severe on the leeward side. In this zone and in the swales between frontal and back dunes, plants that are less resistant to salt spray can survive. They may grow anywhere on the dune but are never long-term residents of the windward side. Some that are common along both northern and southern coasts are seaside-goldenrod *(Solidago sempervirens,* fig. 4.11), dune sandspur *(Cenchrus tribuloides,* fig. 4.12), and groundsel tree *(Baccharis halimifolia,* fig. 4.13).

4.13. Groundsel Tree *(Baccharis halimifolia)*

4.14. Croton *(Croton punctatus)*

4.15. Pennywort *(Hydrocotyle bonariensis)*

4.16. Seaside Evening-Primrose *(Oenothera humifusa)*

4.17. Yaupon *(Ilex vomitoria)*

Seaside-goldenrod is a member of the aster family with fleshy leaves and numerous flower heads, each with eight to twelve golden-yellow ray flowers. It is usually about 3 feet (1 m) tall and sometimes grows on the edges of salt marshes.

Dune sandspur is a grass with seeds enclosed in a spiny bur. An encounter with one of these on a barefoot walk in the sand is sure to bring a cry of pain.

Groundsel tree is a shrubby plant with thick, gray-green leaves and cream-colored flower heads. The female flowers are on separate plants, and in autumn the seeds have prominent white bristles. It sometimes grows on the edges of salt marshes.

The following plants are more common on dunes south of Virginia.

Croton *(Croton punctatus,* fig. 4.14) has whitish-gray, oval leaves, with star-shaped hairs that are visible with a hand lens. The female flowers are on separate plants that bloom from May to November producing a three-lobed seedpod. The leaves have a fragrant odor when crushed.

Pennywort *(Hydrocotyle bonariensis,* fig. 4.15) has bright green, round leaves attached in the center like a small flat umbrella. It also grows on dunes of the Pacific coast.

Seaside evening-primrose *(Oenothera humifusa,* fig. 4.16) has cup-shaped flowers with four yellow petals tinged with pink, and stems that form a mat spreading over the sand.

Arrowleaf morning glory *(Ipomoea sagittata)* grows as a vine trailing over the sand with white, trumpet-shaped flowers.

Yaupon *(Ilex vomitoria,* fig. 4.17) is a shrub or small tree with leathery evergreen leaves and red berries. It is very sensitive to salt spray and often has no leaves or branches on the side facing the ocean. A tea containing caffeine can be made from its leaves.

On the leeward sides of back dunes and on sandy flats not exposed to excessive salt spray, conditions are favorable for several species of woody plants. Two of these are wax-myrtle *(Myrica cerifera)* and bayberry *(M. pensylvanica,* fig. 4.18). Both are shrubs with aromatic leaves and fragrant, gray, wax-covered berries on the female plants. In colonial times, the wax was melted in boiling water

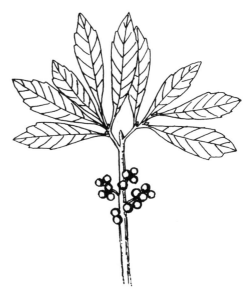

4.18. Bayberry *(Myrica pensylvanica)*

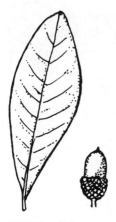
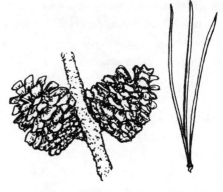

4.20. Pitch Pine *(Pinus rigida)*

4.19. Live Oak
(Quercus virginiana)

and then skimmed to make pleasant-smelling candles. Wax-myrtle leaves are evergreen and usually about $1/2$ inch (1 cm) wide. It grows on beaches from southern New Jersey south to Florida and Texas. Closely related and similar, bayberry ranges from North Carolina to Newfoundland.

Scrub oak or bear oak *(Quercus ilicifolia,* fig. 7.1) grows on sandy shores north of Virginia, and live oak *(Q. virginiana,* fig. 4.19) often occurs on back dunes far from salt spray south of Virginia. All along the coast, pitch pine *(Pinus rigida,* fig. 4.20) may be a large tree or stunted and deformed, depending on the exposure to salt spray.

Dune Succession

On sand dunes along the eastern margins of the Great Lakes, there appear to be successional sequences. The main dune builder and stabilizer is beach-grass, accompanied by other herbaceous plants on the frontal dunes. On the leeward sides and in dune swales, shrubby willows *(Salix* spp.) and cottonwood *(Populus* spp.) are able to grow. After a very long time, these trees are replaced by deciduous forest. In each of the stages, the sandy substrate is modified enough to permit plants of the next stage to become established. Succession on dunes along oceanic coasts is retarded and modified by salt spray. However, when the Pilgrims landed on Cape Cod in 1620, they recorded the dunes as wooded to the shore.

In order for ecological succession to take place, the plants of one stage must be able to modify the environment enough for the survival of plants in the next stage. Some environments, such as the desert and the Arctic tundra, are so harsh that plants cannot modify growing conditions. In these regions there is no succession. Some ecologists studying dune vegetation along the southeastern Atlantic coast of the United States have concluded that succession as it takes place in wetlands and deciduous forests may not occur on coastal sand dunes.

5

Salt Marshes

Salt Marshes and Sand Dunes

Salt marshes and sand dunes are related in several basic ways. First, most of them are located in coastal regions and are often in proximity to one another. Secondly, some of the plants that grow in salt marshes also grow on beaches and sand dunes. In addition, they both have zones of vegetation that are strongly influenced by salt water or salt spray.

Typically on non-rocky shorelines, sand accumulates by wave action and dunes develop above the high-water line. When there are promontories or barrier beaches, salt marshes develop on the landward sides in areas protected from wave action. Examples of this kind of landform can be seen on Cape Cod, Massachusetts, which has sandy beaches on the Atlantic side and salt marshes on the bay side. The barrier islands along the Atlantic coast also have wave action and sand along the seaward sides and salt marshes on the landward or bay sides. In the United States, about 60 percent of the salt marshes are located along the Gulf coast, 31 percent along the Atlantic coast, and 7 percent on the Pacific coast.

Origins and Importance

Salt marshes are very productive and highly valued ecosystems. In the production of biomass (the total amount of biological tissue), salt marshes are more productive than the most fertile corn or wheat fields. More than two-thirds of all shellfish (oysters, clams, crabs, and shrimp) and marine food fish spend part of their lives in salt marshes. But as with other wetlands, salt marshes have been destroyed at an alarming rate by the activities of humans. For example, in the thirteen years between 1950 and 1963, Massachusetts lost 25 percent of its salt marshes; by 1965 Connecticut had lost 50 percent of its salt marshes; and in the seven years from 1964 to 1971, Long Island, New York, lost 3,952 acres (1,600 ha). In the United States as a whole, in the twenty years from 1954 to 1974 about 10 percent of all coastal wetlands were lost.

Salt marshes have developed in the relatively flat areas along the Atlantic, Gulf, and Pacific coasts. In general they form in lagoons and bays on the shoreward sides of barrier beaches and islands, where they are protected from the full force of the waves. From a distance they appear to be extensive grasslands. The most important environmental factors that influence salt marshes are ocean tides. The plants that grow there have a high tolerance for salt and are called halophytes. The amount of salt that can be tolerated varies, with some species surviving higher concentrations than others. While factors such as oxygen and freshwater availability also may exert an influence, zones of vegetation in salt marshes are mainly reflections of differing concentrations of salt in the substrate.

Vegetation Zones

Low marsh. The salt concentration in a salt marsh depends on the frequency and duration of flooding by tidal salt water. The greatest salinity occurs in those areas that are underwater twice daily at high tide but are normally not flooded at low tide. This intertidal zone is on the seaward edge and is called the low marsh. It is typically occupied by a single species, smooth cord-grass *(Spartina alterniflora,* fig. 5.1), a coarse grass that may grow to a height of 6 feet (1.8 m). Since the plant grows in a substrate almost saturated with salt, some of the salt is absorbed by the roots. When salt is present in the ground or in the air as salt spray, it has a deadly effect on most plants. Smooth

5.1. Smooth Cord-Grass *(Spartina alterniflora)*

5.2. Salt-Meadow Grass
(*Spartina patens*)

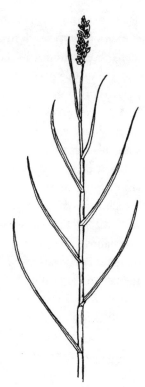

5.3. Spike-Grass (*Distichlis spicata*)

5.4. Marsh Elder (*Iva frutescens*)

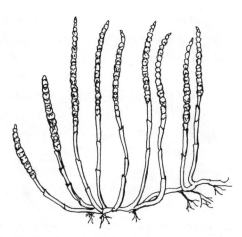

5.5. Perennial Glasswort (*Salicornia virginica*)

cord-grass survives salinity that would kill other plants with special glands that eliminate salt from internal tissues. The result of the action of these glands can be observed as white crystals on stems and leaves before they are washed away by the next high tide.

Over an extended period of time, sediments carried by incoming tides and inflowing streams are trapped by marsh vegetation. This raises the level of the marsh until part of it is out of water except during the highest tides. The accumulation of sediments causes a gradual migration seaward of the zones of vegetation. That part of the marsh that is flooded only during spring tides and storm tides has a substrate with less salinity and is called the high marsh.

Spring tides occur when there is a full moon or a new moon. Low tides or neap tides are at first-quarter and third-quarter moons when the gravitational attractions of the sun and moon are modified because they are at right angles to one another.

High marsh. A characteristic plant of the high marsh is salt-meadow grass *(Spartina patens,* fig. 5.2). In contrast to smooth cord-grass, it is a small, fine grass that is usually no higher than 2 feet (60 cm). The stem has a weak area near its base at which the plant bends. Since it grows in dense stands, when one plant bends it influences adjacent plants to bend also. Eventually, large patches of plants are lying flat on the ground in curled patterns called cowlicks, like hair cowlicks. These swirled patches are so consistent in salt marshes that they seem to be more than accidental, suggesting an as-yet-undiscovered evolutionary survival advantage. Other plants that often grow with salt-meadow grass in the high marsh are spike-grass *(Distichlis spicata,* fig. 5.3) and marsh elder *(Iva frutescens,* fig. 5.4).

Although less salty than the low marsh, the high marsh is still a very salty habitat. As in smooth cord-grass, salt-elimination glands are also present on the leaves and stems of salt-meadow grass. In depressions called pannes, pools may form in which the concentration of salt is increased by evaporation. In some pannes the concentration of salt may be as high as fifty parts per thousand. Normal sea water is about thirty-five parts per thousand. Plant species that are able to grow in or around these depressions are perennial glasswort *(Salicornia virginica,* fig. 5.5), samphire *(S. europaea,* fig. 5.6), southern sea-blite *(Suaeda linearis,* fig. 5.7), saltwort *(Salsola kali,* fig. 6.5), salt-marsh sand-spurry *(Spergularia marina,* fig. 5.8), salt-marsh agalinis *(Agalinis maritima,* fig. 5.9), and the short form of smooth cord-grass.

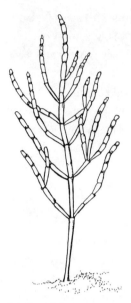

5.6. Samphire
(Salicornia europaea)

5.7. Southern Sea-Blite
(Suaeda linearis)

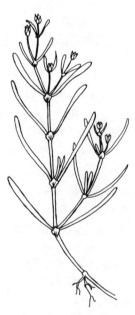

5.8. Salt-Marsh Sand-Spurry *(Spergularia marina)*

5.9. Salt-Marsh Agalinis
(Agalinis maritima)

The latter is a dwarf form of smooth cord-grass that does not get taller than 2 feet (60 cm).

Plants that grow in the pannes may also grow along the upland margins of the salt marsh. Since it is even higher than the high marsh, plants with less tolerance for salt may also grow there. Additional salt-tolerant plants are described below.

Coast-blite (Chenopodium rubrum, fig. 5.10) is a fleshy herbaceous plant with leaves that are mostly alternate in arrangement, oval to triangular in shape, coarsely toothed, and tinged with red at maturity. It grows in high marshes from Nova Scotia to New Jersey.

Groundsel tree (Baccharis halimifolia, fig. 4.13). See chapter 4 for a description of this species.

Spearscale (Atriplex patula, fig. 5.11) is an erect or prostrate herbaceous plant with fleshy leaves that are mostly alternate in arrangement and triangular-to arrow-shaped. It grows on the shoreward margins of the marsh.

Perennial salt-marsh aster (Aster tenuifolius, fig. 2.15) has a smooth

5.10. Coast-Blite *(Chenopodium rubrum)*

5.11. Spearscale *(Atriplex patula)*

5.12. Sea Oxeye (*Borrichia frutescens*)

5.13. Sea Lavender
(*Limonium carolinianum*)

zigzag stem that may be 2 feet (60 cm) high with pale purple flower heads, each over $1/2$ inch (1 cm) wide. It has alternate, narrow leaves. A related species, annual salt-marsh aster (*A. subulatus*), is very similar but has flowers less than $1/2$ inch (1 cm) wide.

Sea oxeye (*Borrichia frutescens*, fig. 5.12) is a shrub that can reach 4 feet (1.2 m) high but is usually smaller, with bright yellow sunflower-like flower heads. It has thick opposite leaves and grows on the high marsh from Virginia to Florida.

Sea lavender (*Limonium carolinianum*, fig. 5.13) is a low herbaceous plant with flowering stems to 2 feet (60 cm) high. It has many tiny pale purple to lavender flowers in a profusely branched cluster at the tip of the stem. Its leaves are lance shaped, widest above the middle, and have long stalks arising from the base of the plant. Sea lavender grows in the high marsh and on the landward edges.

Common reed (*Phragmites australis*, fig. 5.14) is a tall grass, up to 14 feet (4 m) high with hollow stems and long tapering leaves. It has a very dense cluster of flowers a foot (30 cm) or more in length at the tip of the stem. This species grows on the shoreward edge of the salt marsh. It is especially

abundant in areas where the natural vegetation has been disrupted by the activities of humans.
Seaside-goldenrod (Solidago sempervirens, fig. 4.11). See chapter 4 for a description of this species.
Seaside plantain (Plantago maritima, fig. 5.15) has narrow, fleshy leaves and inconspicuous flowers on a stem 8 inches (20 cm) high. It grows on the outer edges of salt marshes, on beaches or dunes, and on the coastal plain. It is circumboreal in distribution and extends along the Atlantic coast south to New Jersey.

Salt Marshes of the Pacific Coast

As the result of a rocky coastline, there are fewer salt marshes on the Pacific coast than on the Atlantic. Where they occur, they have endured the same rate of destruction as those on the Atlantic and Gulf coasts. As in the east they have been dredged, filled, diked, and ditched. In the San Francisco Bay estuary, 95 percent of the marshes have been diked or filled.

The same genera of plants, and in some instances, the same species, grow in the salt marshes of both coasts. The intertidal zones or low marshes of southern California coastal wetlands are occupied exclusively by Pacific cord-grass *(Spartina foliosa)*. Some typical plants of the high marsh and landward edge are spike grass, glasswort *(Salicornia pacifica)*, sea-blite *(Suaeda californica)*, saltwort *(Batis maritima)*, coast-blite, pickerelweed *(Pontederia* spp.), sticky sand-spurry *(Spergularia macrotheca)*, brass buttons *(Cotula coronopifolia)*, and Pacific silverweed *(Potentilla pacifica)*.

5.14. Common Reed *(Phragmites australis)*

5.15. Seaside Plantain *(Plantago maritima)*

6

Through the Year

The Sea and the Intertidal Zone

In temperate regions, life in the ocean, like that on land, responds to the tilt of the earth's axis and the revolution of the earth around the sun. During winter, when the northern hemisphere is tilted away from the sun, the ocean receives less radiation because the days are shorter and the sun is lower in the sky. Even much of the radiant energy that does strike the water is reflected back into space. The lowered amount of energy available for photosynthesis reduces the number of phytoplankton organisms to their lowest level. Since their food supply has dwindled, the zooplankton—and all the animals that depend on them as a source of food—are at their lowest ebb also. The cold waters are whipped by winter storms that send waves crashing into the shorelines, sending salt spray to its inland limits.

In late winter, after months of cooling, cold surface water may sink over the continental shelf and be replaced by warmer, nutrient-rich water from the depths. More importantly, in early spring the radiant energy begins to increase as the days lengthen and the sun rises higher in the sky. The result is an explosion of planktonic algal growth. It has been estimated their numbers may increase as much as sixty thousand times or more.

During summer months, the surface water warms, forming a layer in which there is little mixing with lower colder layers. The number of phytoplankton continues to increase, but by midsummer the mineral nutrients, mainly nitrogen and phosphorus, in the surface water are greatly reduced and growth rates decrease. The great number of zooplankton and other small animals continue to feed on the algae, reducing the number still more. An interesting late summer phenomenon in some temperate areas is the growth of bioluminescent algae of a group called dinoflagellates. When the water around them is disturbed, these organisms produce a greenish-yellow light. A boat traveling through their waters at night leaves a ghostly trail of light.

The surface layer of water over the continental shelf begins to cool in autumn as the days shorten. Eventually the temperature gradient is eliminated and wind disturbances of surface water cause mixing with nutrient-rich water from deeper levels. With the increase in available nutrients comes a resurgence of algal growth. It is brief because by this time the days are growing short and radiant energy for photosynthesis is no longer available.

In the intertidal zone, late fall and winter are times when the fewest macro-algae can be observed. Winter forms of sea lettuce *(Ulva lactuca,* fig. 2.1) are a fraction of the size they are in summer. Broad-leaved kelp or sweet tangle *(Laminaria saccharina,* fig. 1.4) and other species of this genus die in winter, except for a regenerative zone of cells in the stipe. The blades are torn away by winter storms and washed ashore or out to sea. Many species of rockweed *(Fucus* spp.) are perennial and show no changes in winter. As the available radiant energy increases in late winter and early spring, the regeneration process begins. Spores of red and purple laver *(Porphyra leucostica, P. umbilicalis),* which are red algae, germinate and grow into membranous plant bodies very similar to that of sea lettuce. A significant difference is that instead of some shade of green, they are purple-red to reddish-black. Regeneration cells that have overwintered in the stipes of holdfasts of some of the brown algae begin the process of growing new blades.

As spring becomes summer, the longer days and higher temperatures stimulate maximum proliferation of intertidal algae. Blades that have been eaten by animals or damaged by wave action are repaired or replaced by constant growth. Summer is a time for harvesting edible seaweeds when they are at their most highly developed stage of growth. As summer passes into autumn, the days shorten and the cycle begins anew.

Beaches, Dunes, and Salt Marshes

On beaches, dunes, and salt marshes, winter is a time of plant dormancy. This is a chemically induced state that had its beginnings in late summer and early fall. In the trees of the eastern deciduous forest, the shortening days stimulate the manufacture in leaves of abscisic acid and perhaps other growth-inhibiting substances. These are transported to the winter buds before the leaves are shed. The same kind of process probably takes place in perennial herbaceous plants of the seashore. Many herbaceous plants have underground stems called rhizomes with buds from which new plants will grow in the spring. Growth-inhibiting substances are transferred to the rhizomes before the leaves fade in late autumn.

The survival value to the plant of growth-inhibiting substances is that they keep the buds from starting to grow during brief warm spells in winter. Before growth can resume in spring, the growth-inhibiting chemical must be eliminated. Plant scientists are not sure how this is accomplished, but it is known that in many species, in order for the bud to begin growth it must be exposed to an extended period of cold weather. After a period of exposure to cold, plants require increasing temperatures and increasing day lengths to resume growth. When these conditions have been met, the first indication of renewed growth is a swelling of the buds. This is a signal that dormancy is broken, and it marks the beginning of ecological spring.

According to the calendar, winter begins on December 21. In the natural world this date has no significance because it is not reflected in any biological event. By December ecological winter is fully established and plants are in full dormancy. Ecologically, the best marker for the beginning of this season is the shedding or the withering of leaves. On the northeastern seashore this may occur as early as October 31, but as one travels southward it will come later.

Spring begins, by the calendar, on March 21. This, too, is a biologically meaningless date since much of the northeastern seashore is still in the grip of ecological winter. Spring begins with the swelling and bursting of buds, which is an indication that growth inhibitors have been dispelled and winter dormancy is at an end. This may be several weeks after March 21, especially in the northeast. As with ecological winter, the further south one travels, the earlier spring will arrive.

As spring becomes summer, there is a progressive development of seashore vegetation. Plants respond to long days and warm temperatures with maximum vegetative growth. This is a characteristic of ecological summer. A feature of late summer and early fall that has ecological significance is the shortening of the daylight period. As this happens, plants begin to close down summer functions with activities that will prepare them for winter. They are stimulated to initiate the manufacture of growth inhibitors and other changes that will terminate in the cessation of photosynthesis for the season.

Ecological spring, summer, and autumn at the seashore can be recognized by the flowers of species that can be seen in bloom only during those seasons.

Beaches and Sand Dunes

Plants That Flower in April

Broom crowberry (Corema conradii, fig. 6.1) is a low-growing evergreen shrub with staminate and pistillate flowers on separate plants. The flowers do not have petals, but the staminate plants have purple stamens. The flowers occur in clusters at the tips of the stems and branches. The leaves are very narrow and about $1/_4$ inch (6 mm) long. The stems are freely branched, forming dense clumps that may be to 16 inches (40 cm) high.

Since pollen is produced on one plant and the pistil is on a separate one, cross-pollination is assured. The fruit is a nutlike berry, brownish-black, about the size of a pinhead, and usually contains three tiny seeds. These plants are of limited value as food for wildlife and of little economic importance. They are difficult to maintain but are sometimes cultivated as ornamentals or novelties.

Broom crowberry is found on dunes, sandy beaches, and sandy or rocky soils along the Atlantic coast, and rarely inland, from Newfoundland to New Jersey. It attains its greatest height and is most abundant in the northern part of its range.

6.1. Broom Crowberry *(Corema conradii)*

Plants That Flower in May

Eastern blue-eyed grass (Sisyrinchium fuscatum, fig. 6.2) is a low, perennial, grasslike herb. It has blue flowers with yellow centers and six petals, each with a hairlike point at its tip. The stem is flattened, appearing to be branched, with a cluster of flowers at the tip of each branch. The stem is taller than the flattened leaves and may reach a height of 20 inches (50 cm). There are several species of blue-eyed grass growing in a variety of habitats.

Eastern blue-eyed grass may be observed on the landward sides of beaches and along the sandy coastal plain from Massachusetts to Florida and Mississippi.

Beach wormwood (Artemisia stelleriana, fig. 4.8) is a perennial with a creeping underground stem. It has numerous yellowish flower heads at the tips of stems that may reach 28 inches (70 cm) in height. It has pinnately lobed leaves that are woolly white on both sides. It is a native of northeastern Asia and Japan and was introduced into North America for its ornamental white leaves. It has escaped from cultivation and is now firmly established on beaches and dunes from Quebec to Virginia. Beach wormwood is also known as dusty miller.

6.2. Eastern Blue-Eyed Grass *(Sisyrinchium fuscatum)*

The genus name *Artemisia* is from Artemis the Greek goddess of chastity, but the connection between this plant and chastity is unknown.

False heather (Hudsonia tomentosa, fig. 4.9) is a low-growing, evergreen shrub with numerous small yellow flowers that grow at the tips of short branches and open only in sunshine. Each flower has five petals and eight to thirty stamens. The leaves are scalelike, overlapping, covering the stem, and densely covered with white down. The stems are prostrate with dense erect branches forming thick, greenish mats up to 8 inches (20 cm) high and 3 feet (90 cm) wide.

False heather is found on beaches and sand dunes along the Atlantic coast from Labrador to Virginia and North Carolina and inland in sandy habitats. It is also known as beach heath and poverty grass.

A closely related species is golden heather *(H. ericoides)*. It has a similar distribution and a similar growth habit but is less common than false heather. It has needlelike leaves that are less downy and do not overlap to cover the stem.

Prickly pear (Opuntia humifusa, fig. 3.1) is a member of the cactus family with yellow flowers up to 3 inches (7.5 cm) wide and sometimes with reddish centers. The flowers have numerous petals and stamens and produce purplish, egg-shaped fruits about 2 inches (5 cm) long. The leaves are small and inconspicuous, brown, and are shed soon after they form. The stem is made up of green, jointed pads that sprawl on the sand forming large mats. Newly formed pads are sometimes erect, 12 inches (30 cm) high. The pads have tiny clusters of reddish-brown barbed bristles, about 1 inch (2.5 cm) apart. In eastern North America this species is usually spineless but the tiny barbed bristles penetrate the skin and are released from the plant at the slightest touch.

This is a very large genus native to North America. It is more commonly associated with the deserts of the Southwest. The above species is the only widespread cactus east of the Mississippi River. Plants growing west of the Appalachian Mountains sometimes have long spines.

Prickly pear is found on shores and sand dunes from Massachusetts to Florida and Texas.

Plants That Flower in June

Beach pea (Lathyrus maritimus, fig. 4.7) is a perennial with clusters of purple flowers on stalks arising from leaf axils. The fruit is a pealike pod about 2 inches (5 cm) long. The leaves are alternately attached and pinnately compound, with six to twelve leaflets and a tendril at the tip. The stems creep over the sand and may be up to 40 inches (100 cm) long.

This species is a native of North America and the cooler parts of Europe and Asia. At maturity the seedpods split longitudinally into two sections releasing several smooth, round, olive-green seeds. These are somewhat resistant to salt water and under some circumstances may be dispersed by sea water.

Beach pea is found on sandy beaches and dunes along the Atlantic coast from Labrador to New Jersey, and inland on sandy shores of Lakes Champlain and Oneida and the Great Lakes.

Sea-beach sandwort *(Honckenya peploides,* fig. 4.10) is a perennial with a horizontal underground stem that develops roots and new plants at intervals. It has white flowers with five petals, five sepals, and usually ten stamens. Pistils and stamens are on separate flowers. The leaves occur in pairs, are bright yellow-green, fleshy, oval in outline, and up to an inch (2.5 cm) long. The stems are fleshy and soft, with branches spreading over the sand, sometimes forming dense colonies of up to 6 feet (1.8 m) in diameter.

Sea-beach sandwort grows on beaches and sand dunes along the Atlantic coast from Newfoundland to Virginia and North Carolina and the Pacific coast from Alaska to California. It is also know as sea-chickweed and sea purslane.

Sea-rocket *(Cakile edentula,* fig. 4.4) is a fleshy annual with a deeply penetrating root system. It has small pale purple to white flowers with four thick sepals, four petals, and six stamens. All members of the mustard family, of which sea-rocket is a member, have two short and four long stamens. The leaves are thick and alternately attached, with wavy, toothed margins or pinnately lobed. Stems are erect, smooth, often bushy-branched, and up to 16 inches (40 cm) high.

A related species, European sea-rocket *(C. maritima)* is similar but has deeply dissected leaves and seedpods with projecting wings. It is a native of Europe and occurs along the northern Atlantic beaches.

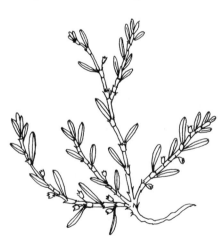

6.3. Seaside knotweed *(Polygonum glaucum)*

Sea-rocket grows on sandy beaches and dunes along the Atlantic coast from Labrador to Florida and inland on sandy beaches along the Great Lakes.

Seaside knotweed *(Polygonum glaucum,* fig. 6.3) is an annual with small pink flowers growing singly or in small clusters in the axils of leaves. The plant is very pale, sometimes nearly white, with fleshy oblong leaves about 1 inch (2.5 cm) long. The stem consists of numerous joints surrounded by prominent silvery membranes. A leaf arises from each joint. The stem is abundantly branched, sprawling on the ground usually with upturned branch tips. The term knotweed in the common name refers to the conspicuous joints or knots where the leaves are attached.

Seaside knotweed grows on sandy beaches and dunes along the Atlantic coast from Maine to Florida and Alabama.

Yellow hedge-hyssop (Gratiola aurea, fig. 6.4) is a perennial with bright yellow, four-lobed, tubular flowers on long stalks from the axils of leaves. The leaves are in pairs, about 1 inch (2.5 cm) long, clasping the stem at their bases. The stem is often creeping at the base with erect branches up to a foot (30 cm) high.

6.4. Yellow Hedge-Hyssop *(Gratiola aurea)*

This plant grows on sandy or muddy shores from Newfoundland and Quebec along the Atlantic coast to Florida and Alabama. Other names for it are golden hedge-hyssop and golden pert.

Plants That Flower in July

Saltwort (Salsola kali, fig. 6.5) is an annual whose green flowers have five sepals and five anthers growing in the axils of leaves. The leaves are alternate, the upper ones slightly under 1 inch (20 mm) long with the midrib prolonged into a spine. The stems are profusely branched and bushy up to 16 inches (40 cm) high. It is wind pollinated and a western variety has been reported as a key hay fever plant in Arizona, Colorado,

6.5. Saltwort *(Salsola kali)*

Oklahoma, and Oregon. The seed is enclosed by an enlarged, winged calyx that aids its dispersal by wind. In addition, the stem breaks at the base and the entire plant rolls in the wind as a tumbleweed, spreading seeds as it goes.

Saltwort is found on sea beaches and dunes along the Atlantic coast from Newfoundland to Louisiana. It also grows along the coast of western Europe.

Seaside spurge (Euphorbia polygonifolia, fig. 4.3) is an annual with small and inconspicuous flowers on stalks arising from cup-shaped structures in the leaf axils. The leaves occur in pairs, are pale green, and are about $^3/_4$ inch (18 mm) long. The stems are red with milky juice, highly branched, and sprawl on the sand forming mats. The seeds and vegetation are eaten by several species of upland game birds, songbirds, and small mammals.

This tiny ground-hugging plant may be overlooked unless one is specifically looking for it. A hardy pioneer of sand dunes, it is capable of surviving in a harsh environment of temperature extremes and shifting sand.

Seaside spurge is found on sandy beaches and dunes along the Atlantic coast from Quebec to Florida and inland around the Great Lakes.

Tall wormwood (Artemisia campestris, fig. 3.9) is a biennial with numerous green nodding flower heads in branched clusters at the ends of stems and branches. The basal leaves are highly dissected, but those on the stem are smaller and less dissected. The stems are very leafy, usually unbranched, and up to 3 feet (1 m) tall.

Tall wormwood produces a dense basal rosette of leaves and a substantial taproot during the first year of growth. A flowering stem appears during the second year, seeds are produced, and the plant dies. Pollination is by wind, and the pollen may be a significant cause of hay fever in some areas.

Tall wormwood grows on beaches and sand dunes along the Atlantic coast from New Brunswick to Florida and inland along the Great Lakes.

Plants That Flower in August

Flat-topped goldenrod (Euthamia tenuifolia, fig. 6.6) has very small flower heads with yellow ray flowers in branched, flat-topped clusters at the tops of stems. The leaves are numerous, very narrow, resin dotted, and fragrant, with clusters of smaller leaves in leaf axils. They are about 3 inches (7.5 cm) long on slender stems that may be 32 inches (80 cm) high. It is a perennial with a creeping, branched, underground stem. The leaves on the lower part of the stem are usually shed early, leaving the upper ones all about the same size.

Flat-topped goldenrod is found in dry or moist sandy soil on shores, beaches, and sand dunes along the Atlantic coast from Nova Scotia to Florida and Texas.

Jointweed (Polygonella articulata, fig. 6.7) is an annual with tiny pink or white flowers in several long clusters at the ends of branches and stems. The leaves are about 1 inch (2.5 cm) long, threadlike, and shed early in autumn. The stems are slender and freely branched, up to 16 inches (40 cm) high.

6.6. Flat-Topped Goldenrod *(Euthamia tenuifolia)*

The tiny nodding flowers of this species are very dainty and, although individually inconspicuous, are usually present in sufficient numbers to make an attractive cluster. The reddish-brown tinged stems often stand out, especially when growing in white sand. Populations of these plants growing along coastal dunes and in blowout areas have prostrate stems and may not exceed 1 foot (30 cm) in length. A form with dark purple-red flowers is observed occasionally.

Jointweed is found in dry sandy soil and on beaches and dunes along the Atlantic coastal plain from Maine to North Carolina and inland along the Great Lakes.

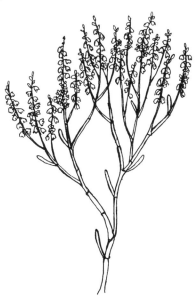

6.7. Jointweed *(Polygonella articulata)*

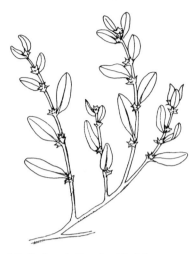

6.8. Seabeach Orache (*Atriplex arenaria*)

Seabeach orache (Atriplex arenaria, fig. 6.8) is a plant that grows anew from seeds every year. This annual has inconspicuous green flowers with stamens and pistils in separate flowers. The leaves are alternately attached, oval-oblong in outline, and silvery in color. The stem may be prostrate or erect, up to 20 inches (50 cm) high.

Seabeach orache grows along shorelines with sandy beaches and dunes from southern New Hampshire to Florida and Texas.

Seaside-Goldenrod (Solidago sempervirens, fig. 4.11) has flower heads with eight to twelve showy yellow ray flowers clustered at the ends of stems and on short branches on the upper parts of stems. The leaves are alternate, fleshy, and numerous, with smooth margins. The stems are erect or arching up to 8 feet (2.4 m) high.

This plant is a perennial with short persistent stem bases and fibrous roots. Since it is self-incompatible, cross-pollination is necessary for seed production and is effected by bees and flies. Seed dispersal by wind is facilitated by a tuft of hairs on each seedlike fruit that serves as a parachute.

Seaside-goldenrod grows on sandy beaches and dunes and on the edges of salt marshes along the Atlantic coast from Newfoundland to Florida, Texas, and Mexico. A less showy form with smaller flower heads and narrower leaves is found south of Virginia.

Salt Marshes and Brackish Water

Plants That Flower in April

Sea oxeye (Borrichia frutescens, fig. 5.12) is a shrub with an underground stem and yellow flower heads. The ray flowers are about thirteen in number, have pistils but no anthers, and are about $1/2$ inch (1 cm) long. The leaves are in pairs, tapered at the base, widest above the middle, and are about 2 inches (5 cm) long. They are covered with fine hairs giving them a whitish color.

The stem has only a few or no branches, and may grow to 3 feet (90 cm) in height.

Sea oxeye grows in salt marshes along the Atlantic coastal plain from Virginia to Florida, Texas, Mexico, and Bermuda.

Plants That Flower in May

Narrow-leaved cattail (Typha angustifolia, fig. 3.6) has tiny inconspicuous flowers densely crowded in a cylindrical cluster at the tip of a long stalk. The staminate flowers are above the pistillate ones, usually with a space of a few centimeters between them. The diameter of the cluster is about $1/_2$ inch (12 mm). The leaves are long and narrow, slightly rounded, and typically no more than $1/_2$ inch (12 mm) wide. The leaves and flower stalk may be 8 feet (2.4 m) high.

This perennial with a thick underground stem grows in North America, Europe, Asia, and Africa. After pollen is shed the staminate flowers wither, but the brown cylinders of seedlike fruits usually persist into the winter months. It eventually disintegrates and the tiny seeds are spread by wind, aided by a parachute of hairs.

Narrow-leaved cattail is found in fresh- and brackish-water swamps from Nova Scotia and Quebec to Florida and Texas.

Plants That Flower in June

Salt-marsh sand-spurry (Spergularia marina, fig. 5.8) is an annual with white or pink flowers in branching clusters at the tips of stems and branches. The flowers have five petals and five larger sepals. The leaves are in pairs, fleshy, very narrow, and about $1^1/_2$ inches (4 cm) long. It has a freely branched stem, erect or nearly prostrate and up to 14 inches (35 cm) long.

This plant is a native in Europe and Asia and was probably introduced accidentally into North America. The seed capsule splits from the top into three sections, releasing several tiny brown wingless seeds. These may be dispersed in mud on the feet of waterfowl, and they are small enough to become wind-borne in a moderate breeze.

Salt-marsh sand-spurry grows in brackish and saltwater marshes from Quebec along the Atlantic coast to Florida and Texas. It is also known as sand-spurry and pink marsh-spurry.

Seaside plantain (Plantago maritima, fig. 5.15) is a perennial with tiny, inconspicuous greenish flowers in a crowded cluster along the upper part of the stem. The flowering stalk is sometimes longer than the leaves. The leaves are narrow and fleshy with one major vein. The leaves may be 6 inches (15 cm) high and the flowering stalk 8 inches (20 cm) high. The very small seeds are probably dispersed in mud on the feet of birds or for short distances by wind.

Seaside plantain has a worldwide distribution north of the temperate zone, and in North America it extends southward in salt marshes along the Atlantic coast to New Jersey.

Spearscale (Atriplex patula, fig. 5.11) is an annual in which pistils and stamens are in separate green flowers in clusters at the tips of upper branches. Lower leaves are usually in pairs. All leaves are mostly triangular with a small lower tooth pointing outward, and with smooth margins. The stems are widely branched, usually erect, but sometimes prostrate, up to 3 feet (90 cm) long.

This is a highly variable plant of North America, Europe, and Asia. It is wind pollinated but usually does not grow in sufficient numbers to be a significant hay fever plant. The hard shiny black seeds are spread by birds and mammals that feed on the plant.

Spearscale is found in salt marshes along the Atlantic coast from New Brunswick to Florida and inland in salty areas throughout most of southern Canada and the United States.

Plants That Flower in July

Coast-blite (Chenopodium rubrum, fig. 5.10) is an annual with tiny, reddish flowers in dense elongate clusters arising from the axils of leaves. The seeds are dark glossy brown and lens shaped. Its leaves are mostly alternate and reddish; the larger ones are oval to triangular with one or more conspicuous teeth on each side, tapering at the base. The stems may be erect or prostrate, 32 inches (80 cm) long, and often branched from the base.

This plant is native to North America, Europe, and Asia. Pollination is by wind, but it is not known to be a significant hay fever plant. A single plant may produce as many as 175,000 seeds. The plant persists until late in the season, and the seeds are relished by many species of song birds.

Coast-blite grows in salt marshes along the Atlantic coast from Newfoundland and Nova Scotia to New Jersey, and inland on salty soil.

Salt-marsh agalinis (Agalinis maritima, fig. 5.9) has purple-pink, funnel-shaped flowers with five lobes. It has fleshy, narrow leaves in pairs, 1$^1/_2$ inches (4 cm) long. The stem is erect and usually no more than 14 inches (35 cm) high, and very often less. The small brown seeds are probably dispersed in mud on the feet of aquatic birds.

Salt-marsh agalinis is an annual found in salt marshes along the Atlantic coast from Nova Scotia and Maine to Florida, Texas, and Mexico.

Sea lavender (Limonium carolinianum, fig. 5.13) is a perennial with small pale purple or lavender flowers in a profusely branched cluster at the tip of a leafless stem. It has five petals and a hairy, five-lobed calyx. The leaves are lance shaped, widest above the middle, and tapering at the base on long stalks of up to 10 inches (25 cm). The stem is erect, arising from thick woody roots, sometimes up to 2 feet (60 cm) high.

Each of the many flowers of this plant produces a single dark brown, shiny, faintly ridged seed enclosed by a persistent calyx. This genus *(Limonium)* consists primarily of desert and salt-marsh inhabitants. It has a worldwide distribution and includes species that are grown in rock gardens and greenhouses and are used in dried flower arrangements.

Sea lavender is found in salt marshes along the Atlantic coast from Labrador to Florida, Texas, and Mexico. It is also known as marsh rosemary.

Plants That Flower in August

Common reed (Phragmites australis, fig. 5.14) is a perennial with small inconspicuous flowers in dense clusters, 8 to 16 inches (20–40 cm) long, at the tips of stems that grow to 15 feet (4.5 m) in height. The flower clusters are purplish at the time of blooming and then become light brown and feathery, darkening somewhat throughout the year. It has leaves of up to 24 inches (60 cm) long and 2 inches (5 cm) wide, alternately attached on opposite sides of the stem. It is able to spread rapidly by fast growing horizontal stems that often grow along the surface of the ground. The stems will sometimes grow over bare rocks for a distance of up to 25 feet (7.5 m).

In some parts of coastal New England, tide gates were constructed to control the daily tidal flooding. In those areas, the salt-marsh grasses were replaced by common reed. In pre-gate days, it could not survive on the shore because it is intolerant of the daily fluctuations in water level caused by the tides.

Common reed is widely distributed in North America. It grows in fresh- and brackish-water swamps and marshes along the Atlantic coast from Newfoundland to Florida and Texas.

Perennial glasswort (Salicornia virginica, fig. 5.5) has tiny inconspicuous green or colorless flowers embedded in the upper joints of stems. There are usually three in a cluster, all at about the same level. Its leaves are reduced to minute pairs of scales. The main stems are prostrate, rooting at joints, forming mats with erect green flower-bearing branches of up to 1 foot (30 cm) high. The flower-bearing shoots are unbranched, or only slightly so.

The genus name *Salicornia* is derived from Latin words that mean "salthorn," alluding to the saline habitat of these plants and their hornlike branches.

Glasswort is found on the margins of salt marshes along the Atlantic coast from southern New Hampshire to Florida and Texas. It is also know as perennial saltwort, woody glasswort, and lead glass.

Samphire (Salicornia europaea, fig. 5.6) is a shallow-rooted annual with tiny, inconspicuous, green or colorless flowers embedded in upper joints of stems. There are usually three flowers in a cluster with the center one attached above and extending beyond the other two. Its leaves consist of pairs of tiny scales. The stems are erect, freely branched, green, fleshy, and jointed. They are salty to the taste when crushed, and may grow up to 16 inches (40 cm) in height.

In autumn the stems turn red, yellow, or orange and are eaten by several species of waterfowl, especially Canada goose and pintail duck. These probably serve as agents of seed dispersal.

This species normally grows just above the high tide level, but it may be flooded during monthly spring tides. It is the most common species of the genus, and it is occasionally observed in the waste areas of pickle factories and salt works.

Samphire and perennial glasswort often occur together in the margins of salt marshes along the Atlantic coast from Quebec to Florida and Texas, but perennial glasswort does not extend inland.

Marsh elder (Iva frutescens, fig. 5.4) is a perennial shrub with greenish, nodding flower heads in the axils of leaves at the ends of stems and branches. It has somewhat fleshy leaves attached in pairs except for upper small ones. The leaves are elliptical with pointed tips and toothed margins, the main ones 4 inches (10 cm) long. This plant has a woody base and may be 10 feet (3 m) high.

Marsh elder grows in salt marshes and their margins along the Atlantic coast from Nova Scotia to Florida and Texas. It is also known as maritime marsh elder and high-water shrub. It is closely related to sea-elder *(Iva imbricata,* fig. 4.6), which grows on beaches and sand dunes.

Perennial salt-marsh aster (Aster tenuifolius, fig. 2.15) has flower heads about an inch (2.5 cm) wide, not numerous, with pale purple ray flowers. It has alternately attached, fleshy, very narrow leaves up to 6 inches (15 cm) long. The stem, up to 2 feet (60 cm) high, has few to no branches and often zigzags between leaves. It is a perennial with fibrous roots and creeping surface runners. Each flower head produces numerous gray, slightly hairy, seedlike fruits with parachutes of reddish-brown hairs. Forms with whitish ray flowers are occasionally observed. This small, attractive aster often goes unnoticed among the tall salt-marsh grasses with which it grows.

A related species, annual salt-marsh aster *(Aster subulatus),* has many smaller flower heads no more than $1/_2$ inch (13 mm) wide with very short ray flowers. It has leaves up to 8 inches (20 cm) long and reaches a height of 3 feet (90 cm). This plant often occurs in the same kind of habitat as perennial salt-marsh aster but has a broader geographic range.

Perennial salt-marsh aster grows in brackish water and salt marshes along the Atlantic coast from Massachusetts to Florida and Louisiana.

Salt-marsh fleabane (Pluchea odorata, fig. 6.9) has pink or purple flower heads with no ray flowers in flat-topped clusters at the tip of the stem. Its leaves are alternately attached, numerous, lance shaped or oval with

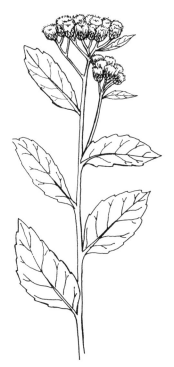

6.9. Salt-Marsh Fleabane *(Pluchea odorata)*

small marginal teeth, up to 6 inches (15 cm) long. The stems are stiff and sticky, up to 3 feet (90 cm) high. This plant is an annual with a faint camphorlike odor. The inner flowers of each flower head produce pollen but have nonfunctioning pistils while the outer flowers produce no pollen and have fertile pistils. This is an adaptation to reduce the possibility of self-pollination. Each of the seedlike fruits has a parachute of hairs that aids in dispersal by wind.

All of the species of this genus in eastern North America are similar in having pinkish flower heads, an odor of camphor, and a tolerance for salt or brackish water. The above species is a colorful contributor to the autumnal aspect of the salt marsh. Its showy flower heads are often used in dried flower decorations.

Salt-marsh fleabane grows in brackish and saltwater marshes along the Atlantic coast from Maine to Florida, Mexico, the West Indies, and inland areas of high salt concentration.

Southern sea-blite (Suaeda linearis, fig. 5.7) has green flowers in clusters in the axils of leaves. The flowers have five stamens, no petals, and a five-lobed calyx that completely encloses the fruit at maturity. Its leaves are very narrow, fleshy, and taste salty when crushed. They are about 2 inches (5 cm) long, have a rounded lower surface, and are progressively shorter toward the top of the plant. The stems are erect, freely branched, and may attain 3 feet (90 cm) in height.

This plant is an annual of eastern North America and the West Indies. Pollination is by wind, but it is not usually considered a consequential contributor to hay fever. Each flower produces a single smooth, glossy, black seed. The whole plant often breaks at the base and rolls with the wind like a tumbleweed, spreading its seeds as it goes.

A related species, white sea-blite *(S. maritima)* is similar, but the upper three lobes of the calyx do not have a prominent keel. It is much smaller, often with prostrate stems, and is found in Europe and along the Atlantic coast from Quebec to New Jersey.

Southern sea-blite grows in salt marshes along the Atlantic coast from Maine to Florida and Texas. It is also known as tall sea-blite.

Swamp rose-mallow (Hibiscus moscheutos, fig. 3.4) has pink flowers up to 6 inches (15 cm) wide, usually borne singly in leaf axils on the upper part of the stem. Its flower has five petals, numerous stamens united into a tube around the style, and a five-lobed calyx. The leaves are alternate, broadly rounded at the base, pointed at the tip with a toothed margin, and up to 8 inches (20 cm) long. The stems are hairy on the upper parts, and grow to $6^1/_2$ feet (2 m) high.

This is a perennial plant with a dense cluster of fibrous roots. The spiny pollen grains stick together in clumps that are transferred from one flower to another by hummingbirds or bees. The seed capsule splits into five sections, releasing several dark, oval, warty seeds. The seed capsule may persist on the dead stems throughout the winter months.

Swamp rose-mallow is found in freshwater and brackish marshes along the Atlantic coast and inland from Massachusetts to Florida and Texas. It is also known as sea-hollyhock.

7

Plants of Special Interest

Early humans were much more familiar with the plants around them than is the average person today. Wild plants made up a large portion of their daily food. Medicines administered by shamans, witch doctors, or medicine men came mainly from plants. Religious rituals were often accompanied by the consumption of hallucinogenic plants. Only the medicine man needed to know the healing and ceremonial plants, but the knowledge of what could and could not be eaten was, of necessity, more widespread. Few modern Americans would survive if their lives depended on finding wild food and medicinal plants. This chapter provides brief discussions, with drawings, of some poisonous, medicinal, and edible wild plants that may be seen at the seashore.

Poisonous Plants

Poisonous plants are those containing substances that have harmful affects on the body if they are eaten or come into contact with the skin. Considering that there are at least 300,000 species of plants, the percentage of known poisonous ones is relatively small. Nevertheless, each year hundreds of cases of poisoning are reported to Poison Control Centers in the United States and Canada. Most cases are children who nibble on poisonous house plants or sample the plant fare in their back yards. Poisoning in adults most often results from misidentifying a poisonous plant and using it for food or an herbal remedy.

There are no reliable physical characteristics that can be used to distinguish poisonous from nonpoisonous plants. Some writers have suggested that plants with red or white berries, milky sap, or an unpleasant odor should be avoided as potentially poisonous. To be sure, there are poisonous plants with these features, but other toxic plants have blue berries; orange, red, or colorless sap; or the pleasant odor of parsnips. To make matters worse, there are harmless and even edible wild plants with these same characteristics.

A belief held by many is that if one can observe other animals eating a plant it is safe for human consumption. This can be a fatal assumption. Although many species of birds eat the berries of poison ivy without apparent harm, it would be very dangerous for humans to consume even one. Therefore, it is not a safe practice to rely on generalities for the identification of poisonous plants. If they are being collected for human use, either as food or for home remedies, any plant that is unknown to the collector should be left where it stands.

Plant Poisons

Functions of plant poisons. Biologists who study plant evolution are interested in determining the origin of the physical and chemical traits of plants. Most characteristics have evolved in response to environmental conditions and contribute to the survival ability of the species. There is still a lot to be learned about why plants produce poisonous substances, but there are several possibilities. One is that they are waste products of metabolism. Since plants do not have excretory systems, wastes cannot be eliminated, as in animals, but must be stored in some part of the plant. Another possibility is that the poisonous substances are compounds that are essential in the normal metabolism and maintenance of the plant and their toxicity to humans and other animals is a coincidence. A third possibility is that poisonous substances have evolved as defense mechanism against plants' greatest natural enemies, insects.

Types of plant poisons. The alkaloids are an important and widespread group of plant poisons. These are compounds that contain nitrogen and react chemically as bases rather than acids. They are almost always bitter tasting and may be present in up to 40 percent of all plant families. Most alkaloids produce a strong reaction in the nervous system when ingested by animals, including humans. This action makes some of them highly toxic, but some are also very important medicinally. The names of alkaloids always end in -*ine* or -*in*, and they are often named for their plant source: caffeine comes from *Coffea arabica*, nicotine from *Nicotiana tabacum*, and cocaine from the coca plant, *Erythroxylon coca*.

The glycosides are a group of poisons that are even more widespread than alkaloids in the plant kingdom. Chemically, glycosides consist of at least one molecule of sugar combined with one or more nonsugar molecules. Although many glycosides are not poisonous, some are lethal. For example, cyanogenic glycosides are broken down by digestive enzymes to release deadly cyanide, which inhibits oxygen uptake by the body cells. Plants with high concentrations of cyanogenic glycosides occur in the rose and bean families. Cardiac glycosides are another group of toxic substances that act directly on the heart muscle.

Other kinds of poisonous substances in plants are oxalic acid and oxalates, phenols, polypeptides, resins, and poisons accumulated from minerals in the soil. For more information on these and other plant poisons, see Kingsbury (1964), Kinghorn (1979), and Harden and Arena (1974).

Types of Reactions to Plant Poisons

Allergies

It has been estimated that at least 15 million Americans are allergic to fungal spores and pollen and thus suffer from hay fever each year. Fungal spores can be in the air at almost any time during the growing season and even during winter months in southern states. Pollen causes discomfort to the allergic during three periods of the year. The first is early spring when wind-pollinated trees release pollen. This is usually the least severe of the allergenic periods. Trees growing on inland dunes or the landward edge of the seashore that may release pollen at this time include scrub oak or bear oak (*Quercus ilicifolia*, fig. 7.1), live oak (*Q. virginiana*, fig. 4.19), pitch pine (*Pinus rigida*, fig. 4.20), and pond pine (*P. serotina*, fig. 3.10).

7.1. Scrub Oak (*Quercus ilicifolia*)

The next in importance of the allergy seasons is the period that occurs in early to mid-summer and is associated with the flowering time of the grasses (family Poacea). The flowers of grasses are so inconspicuous that most people would not even recognize them as flowers. Roses, however, are very conspicuous and bloom at about the same time as the grasses. For this reason, allergic reactions at this time of year are sometimes incorrectly referred to as "rose fever." This is a misnomer because roses are pollinated by insects and very few of their pollen grains get into the air to cause allergic reactions.

The most severe hay fever season, and the one that affects the greatest number of people, is in August and September. This is when ragweed blooms. Although common ragweed (*Ambrosia artemisiifolia*) is not considered a plant of beaches, sand dunes, or salt marshes, it is very widespread, and it would not be unusual to find it growing on the landward side of the seashore along the Atlantic coast where it is not exposed to salt spray.

Skin Irritations

The plant that most commonly causes skin irritations or dermatitis in North America is poison ivy. This plant contains a substance known as urushiol, an oily resin, to which most people are allergic. This toxin is a colorless or milky fluid within special canals in all parts of the plant except the pollen. At least 2 million people each year develop skin rashes from exposure to this compound.

For individuals sensitive to the toxin it might be helpful to review several facts. (1) You cannot get a reaction from simply touching a leaf or stem. The plant part must be bruised or broken so that the canals are ruptured and the toxin comes in contact with the skin. (2) Dead leaves and stems will cause a reaction as readily as green ones. (3) The plant should never be burned because the vaporized toxin and particles in the smoke may affect the eyes, nose, and lungs.

A traditional remedy for exposure to poison ivy or poison oak is to wash with strong soap as soon as possible after contact. But according to the *American Medical Association Handbook of Poisonous and Injurious Plants*, this is not a good idea. It only takes about ten minutes for the toxin to penetrate the skin. If a strong soap is used, the natural body oils will be removed and any remaining toxin may penetrate even faster. Since urushiol is not soluble in water, the recommended treatment is to wash in plain running water without soap.

A common misconception is that the fluid from the blisters of the dermatitis will spread the rash. When the toxin penetrates the skin, it combines chemically with deeper skin tissues. All of the toxin interacts with the cells and thus the greater the exposure, the more severe the rash. Since all of the toxin undergoes an irreversible chemical change once it penetrates, there is none left in the fluid of the blisters, so the rash cannot spread when the blisters burst. There are commercial lotions that claim to protect the user from the urushiol toxin. For those who are allergic, though, the best practice is to learn to recognize the plants and stay away from them.

Poisonous Plants in the Field

Algae

Most of the algae that make up phytoplankton are harmless, but a few species, in a group called dinoflagellates, produce substances that are very toxic to humans. These organisms are normally present in very low numbers, but under some circumstances they will grow to as many as 2 million per quart, giving the water a distinct reddish color. This is called the red tide, and it is not known exactly what triggers the sudden increases or blooms. Sometimes they are associated with excessive rainfall and the accompanying runoff of large quantities of nutrient-rich water. The blooms usually occur in the warm months from April to September. In the Gulf of Mexico they have been linked with high salinity. Some scientists have noted that the number of red tide blooms has increased as ocean pollution has increased.

Red tide organisms produce more than one type of toxin. One type is not poisonous to shellfish but will accumulate in their tissue as they feed on the organisms. Humans eating shellfish such as oysters, clams, and mussels may be afflicted with paralytic shellfish poisoning (PSP) or neurotoxic shellfish poisoning (NSP), depending on the kind and quantity of toxin ingested. The poisons that cause these conditions are so strong that if the nerves controlling the diaphragm and breathing are affected, paralysis and death may follow. Even the salt spray from red tide water may cause skin and eye irritation or coughing and wheezing. Since these toxins are not changed by heat, cooking the shellfish does not deactivate the poison. Although there have been incidents of paralytic shellfish poisoning such as one in Massachusetts in 1972, where thirty people became ill, to date there have been no known deaths.

A more common sickness from seafood is ciguatera poisoning. It is caused by eating fish carrying a toxin that is believed to have been produced by one or more species of dinoflagellates. At least four hundred species of marine fish have been found to be infected with the poison. There have been notable outbreaks of ciguatera poisoning in Florida, Hawaii, Virgin Islands, and Puerto Rico. It is a worldwide problem, and more that 80 percent of the adult residents of the United States have reported at least one incident of poisoning. Symptoms of ciguatera poisoning are headache, vomiting, stomach cramps, irregular heartbeat, and, in severe cases, hallucinations and death.

Fungi

Of all the members of the plant and fungi kingdoms, the mushrooms are the most notorious as poisoners. This reputation is justified for some species, but as with green plants, only a small percentage of the total number of mushroom species are known to be toxic. There are several thousand species of mushrooms in the United States. It is estimated that about a hundred species bring about harmful reactions when ingested, and no more than ten are deadly poisonous. Although all species have not been tested for toxicity, it is believed that these estimates are not likely to change. In folklore, poisonous mushrooms are sometimes referred to as toadstools. The word is derived from a German word that means death's stool. It is not a scientific term.

As with poisonous green plants, there are no rules that a novice can follow for identifying poisonous mushrooms. The differences between edible and poisonous species are often so slight that it requires a trained expert to tell them apart. In addition, a single cluster of mushrooms may include edible and poisonous species growing side by side. This does not seem to deter persistent mushroom gatherers. Thus there are hundreds of cases each year of mushroom poisoning and two or three fatalities. The wild mushroom enthusiast should keep in mind a bit of pithy folk wisdom:

> There are old mushroom hunters
> There are bold mushroom hunters
> But there are no old, bold mushroom hunters.

There is great variation in the ways humans react to mushroom poisoning. No single set of symptoms can be associated with all cases. Species that cause a reaction in some people are eaten by others without problems. The degree of toxicity of a species may depend on the season it is collected, the geographic area in which it grows, or the health of the consumer. Unlike many of the poisonous green plants, poisonous mushrooms apparently do not have a bad odor or a bitter taste. A survivor of poisoning by one of the most deadly mushrooms reported that it was delicious.

Lichens

Most lichens are harmless but a few species in the Midwest and northwest are known to have been responsible for the poisoning of livestock. At least one poisonous compound found in lichen is usnic acid. The effect of the poison is mild to severe paralysis.

Ferns

Bracken fern (Pteridium aquilinum, fig. 7.2) grows in open woodlands, woodland borders, open fields, and along the inland margins of sandy shorelines. It has brown stems, fronds with three parts, and may grow to be several feet in height. The various forms of this species are widespread throughout the northern hemisphere. It contains several poisonous substances including one that destroys thiamin (vitamin B_1) and at least two that are carcinogenic (cancer-causing substances). There have been serious cases of poisoning in cattle and horses, and the carcinogens can be transmitted to humans in milk. Although it has not been proven, eating bracken fern fiddleheads is probably a risk to human health and is not recommended.

Gymnosperms or Conifers

Pine. There are numerous herbal remedies in which pine needles and resin are used to prepare a

7.2. Bracken Fern *(Pteridium aquilinum)*

tea. Most of these are harmless, and some may even be therapeutic. However, some pines are known to have harmful effects on humans and other animals. In the southeastern pine forest, loblolly pine *(P. taeda,* fig. 7.3) has caused death to cattle that have eaten the needles. This tree is a rapidly growing one that may be 90 feet (27 m) or more in height. It has rough, yellow to red-brown twigs and needles in clusters of threes. The needles are 6 to 9 inches (15–23 cm) long, yellow-green in color, and remain on the tree for about three years. Loblolly pine has the widest range of habitats of all the southern pines. It grows along the Atlantic and Gulf coastal plains from New Jersey to Texas.

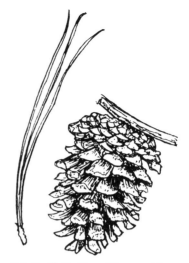

7.3. Loblolly Pine *(Pinus taeda)*

Angiosperms or Flowering Plants

Poison ivy (Toxicodendron radicans, fig. 7.4) is a climbing or creeping woody vine. It has inconspicuous, whitish, hanging clusters of flowers that become clusters of white, berrylike fruits. It has alternate compound leaves with three leaflets. The leaves may be reddish when they first appear, become shiny when mature, then often turn red in autumn. This plant has several varieties that occupy a wide range of habitats. It is the best known and most widely distributed species of a group of related plants that includes poison oak *(T. pubescens)* and poison sumac *(T. vernix).* They all contain the same poisonous substance.

Seaside spurge (Euphorbia polygonifolia, fig. 4.3) is a small

7.4. Poison Ivy *(Toxicodendron radicans)*

sprawling plant of beaches and sand dunes. It is one of a very large genus of plants that includes at least sixteen hundred species. Among these are herbs, shrubs, trees, and cactus-like succulents. Most of them have inconspicuous, greenish flowers similar to seaside spurge. In some species, the flowers are surrounded by colorful leaf-like or petal-like bracts that make them attractive as ornamental houseplants. Crown-of-thorns *(E. splendens)* and poinsettia *(E. pulcherrima)* are examples. Like seaside spurge, many species of this group contain a milky sap that can be seen when a leaf is removed or a stem is crushed. The juice is highly toxic if ingested, and, on the skin of sensitive individuals, it may cause rashes and blisters.

Medicinal Plants

Their Importance

The shamans and medicine men of primitive cultures were probably the first professional men. Since most of the medicines they dispensed came from plants, they were, of necessity, botanists. This strong bond with plants by those that practice the healing arts has been a characteristic of human societies from prehistoric times. It continued into modern times until near the end of the 1800s. Even at that date, many medical doctors were botanists and most professional botanists were physicians. In 1900 about 80 percent of the drugs prescribed by physicians came directly from plants. The development of organic chemistry, beginning at about that time, initiated an era of synthetic medicines. Although the development of synthetic drugs has continued into the present, 35 to 40 percent of all prescribed drugs are still either natural plant compounds or plant compounds in combination with synthetic substances.

Thus plants are still as important in the practice of medicine today as they were to the ancient shamans and medicine men. Today they serve the medical profession in at least three ways. First, almost 25 percent of the drugs prescribed by modern physicians come directly from plants. Second, some plant compounds are used as essential components in the manufacture of medicinal drugs. Third, natural plant drugs may serve as models for the synthesis of identical or similar substances.

In herbal medicine, the ability of willow bark to alleviate pain and reduce fever has been known for two thousand years or more. This effect is mainly the result of a glycoside, salacin, which was isolated and identified in the nineteenth century. Salacin can be converted into salicylic acid. In 1899 salicylic acid was combined with acetic acid to form acetylsalicylic acid. This substance was given the name aspirin and is probably the most widely used medicine in the world.

Only a very small percentage of the known species of plants have been chemically analyzed for medicinal drugs. One can only speculate as to the value of the medicines that remain to be discovered. It is clearly of great importance to maintain habitats for the survival of wild plants including such areas as national, state, and municipal parks, wildlife preserves, and wilderness areas.

Herbal Medicine

In the early days of colonization in North America, physicians and hospitals were few or nonexistent. The settlers had no choice but to rely on herbal medicine for treating illness and injuries. They had brought with them a rich heritage of herbal remedies from Europe, and they soon added to these by including treatments learned from Native Americans. The result is that there are folk remedies for almost every ailment experienced by humans.

A complete description of all the plants and the medicinal uses that have been made of each would require a very large book. Only a few of these remedies have been subjected to controlled testing to verify their effectiveness. Some treatments can be traced to the ill-conceived doctrine of signatures, which held that the shape of a leaf, root, or seed, if it resembled a human organ, determined its use in healing. Other herbal remedies can be traced to a time when magic and mysticism were associated with certain plants. Some folk remedies do make use of plants that contain powerful medicinal drugs—some so powerful, in fact, that a misjudged dose could result in death. It is not surprising, then, that most modern physicians view herbal medicine as little more than quackery.

There is an abundance of justification for this attitude, but to disregard all herbal medicine runs the risk of throwing out the baby with the bathwater. Modern medicine has its roots in folk medicine and there may be information that it can still impart.

Poisonous plants are often components of herbal remedies. Sometimes the only feature in the use of a plant that distinguishes it as medicinal or poisonous is the size of the dose. Experienced practitioners of herbal medicine can usually recognize the signs of acute poisoning, but subtle symptoms from repeated exposure to small doses of a toxic plant drug may not be so easily recognized, even by an experienced herbalist. Modern laboratory techniques are usually necessary to detect damage to internal organs such as the liver or kidneys. However, in many parts of the world, especially in underdeveloped countries, herbal medicine is the chief source of treatment for all human ailments. A study conducted by the World Health Organization concluded that the only way developing countries can achieve minimum health needs is to make use of traditional folk medicine.

It is possible there are effective folk remedies needing only to be clinically tested. It is probable, too, that there are some that should be discontinued. Modern research is constantly providing new information with which to evaluate herbal remedies. Below is a list of a representative group of plants from seashore habitats that have been used in herbal medicine. The descriptions are offered for their historical interest only and not as recommendations for home remedies. Unless the user is thoroughly familiar with the plant components, most herbal remedies are inadvisable. For the inexperienced herbalist, the local drugstore can provide better and safer medicine.

7.5. Horse-Mint (*Monarda punctata*)

Horse-mint (Monarda punctata, fig. 7.5) is a perennial with two-lipped, pale yellow flowers spotted with purple. The flowers are in dense whorls in the axils of upper leaves. Its leaves are in pairs, lance shaped, and 1 to 3 inches (2–8 cm) long with shallow teeth. The stem is unbranched and 1 to 3 feet (30–90 cm) high.

As with other species of this genus, the leaves can be used to make a fragrant tea. Native Americans and early settlers used these plants for a variety of ailments including upset stomach, intestinal gas, headache, colds, sore throat, and bronchial problems. Externally, the tea was used for acne and other skin problems.

Horse-mint has a wide distribution, but one variety is common along the Atlantic coastal plain from New Jersey to Florida and Texas.

Horseweed (Conyza canadensis, fig. 7.6) has tiny, white to pink flower heads with vertical ray flowers on numerous branches from the axils of upper leaves. Its numerous narrow leaves are attached alternately and have mostly smooth margins. The stem is erect, unbranched at the base, covered with stiff hairs, and grows up to 6$^1/_2$ feet (2 m) high.

The whole plant steeped in water has been recommended as a tonic, a diuretic, and a treatment for diarrhea and internal bleeding. Some tribes of Native Americans used a tea made from the roots for menstrual irregularities and vaginal discharges. The leaves are known to cause contact dermatitis in sensitive individuals.

Horseweed has a wide distribution in North America, but one variety grows chiefly in coastal dunes or along the Atlantic and Gulf coastal plains from New England to Florida and Texas.

7.6. Horseweed *(Conyza canadensis)*

Prickly pear (Opuntia humifusa, fig. 3.1) is described in chapter 6. The plants of this genus have been used in home remedies for a variety of aliments. The flat stems have been used as poultices on bruises and on the breasts of mothers to increase the flow of milk. The flowers have been used to make a tea for kidney ailments, and the roots have used in a treatment for dysentery.

Sea lavender (Limonium carolinianum, fig. 5.13) is described in chapter 6. The root has a very bitter and astringent taste, and in the nineteenth century was used to treat diarrhea and dysentery in eastern North America. An ointment made from the powdered root has been used as a soothing treatment for hemorrhoids, chronic gonorrhea, and inflammation of body orifices.

Seaside-goldenrod (Solidago sempervirens, fig. 4.11) is described in chapter 6. Several species of this genus have been used medicinally. Seaside-goldenrod is reported to be useful in the treatment of a variety of ailments including kidney stones, hemorrhaging, diphtheria, dysentery, and wounds. In the Ozark Mountains, the roots of some species are chewed as a treatment for toothache. Some species may cause allergic reactions on contact with sensitive individuals. The flowers of this genus are a source of a yellow or gold dye for wool.

Spearscale (*Atriplex patula*, fig. 5.11) is described in chapter 6. The newly ripened seeds of this species, bruised and soaked in alcohol for six weeks, produce a tincture that is reported to be a mild laxative, a cure for headaches, and a treatment for the first attacks of rheumatism.

Edible Wild Plants

In hunter-gatherer societies, humans were essentially vegetarians. This was especially true before the invention of the spear and the bow and arrow. Even after these tools were invented, plants still made up most of the human diet. The species that were used for food by the hunter-gatherers are still growing in those areas, but very few, if any, are important sources of human food today. Instead, the cereal grains—wheat, rice, corn, oats, barley, and millet—are the main food plants in the modern world.

The edible native plants in any region of the earth are better adapted for survival in the climatic conditions there and are sometimes more nutritious than imported cereal grains. Thus the cultivation of native wild food plants may offer a partial solution to escalating local food shortages. Exploration of this possibility is an appropriate direction for agriculture of the future.

Why Know Them?

In this high-tech era of rapid transportation and very efficient freezers, when well-stocked supermarkets are available to almost everyone, who needs to know about edible wild plants? For the purposes of survival, probably no one. Even if all transportation and freezing facilities failed and supermarkets had empty shelves, knowledge of edible wild plants would be of little value to residents of New York City, Philadelphia, Chicago, or Los Angeles. There simply are not enough edible wild plants out there to feed so many people. In the event of total failure of the supply system and electricity, millions would starve to death. When hunter-gatherers foraged for edible plants, the population was measured in tens of individuals per hundreds of square miles rather than in the millions.

The most valid reason for learning to identify edible wild plants is probably the same reason that people climb mountains: because they are there. Those who love the outdoors find satisfaction in being able to recognize poisonous, medicinal, and edible plants. On the practical side, it is always possible that a camper or hiker could become lost in the wilds. Knowing some edible plants could be very helpful. In addition, being able occasionally to prepare a complete meal with wild plants is a novelty that is sure to surprise and perhaps delight dinner guests. Euell Gibbons (1966) has described elaborate wild food dinners in his home at which every dish was a conversation piece.

Plant Conservation

It is especially important to consider plant conservation when discussing the collection of wild food plants. The individual plants of a given species are seldom randomly distributed throughout their geographic range. Instead, they often occur in scattered clumps in those parts of the range where environmental conditions are suitable for their growth and reproduction. In collecting enough plants for a single meal, an entire local colony could be eliminated. This is less damaging when the plants are perennials and are cut at ground level leaving the rootstock to generate new plants. If the plants are annuals, it is more damaging because young shoots, before they produce flowers and seeds, are usually the most desirable for food.

The edible plants of beaches, sand dunes, and salt marshes listed below are for historical interest only. These habitats are fragile and the plants growing there are usually not present in great abundance. They are never sufficiently abundant to support the pressure of extensive food gathering.

In many regions of the world, especially island cultures, seaweeds have been important sources of food for thousands of years. They are today grown as crops, and harvested from underwater plantations in some areas. There is no known record of a marine plant species threatened with extinction because of food gathering. They are more likely to be endangered by ocean pollution. Seaweeds have never been accepted as a common source of food in North America, but extracts from them are widely used.

Edible Plants in the Field

The Intertidal Zone

Dulse (Palmaria palmata, fig. 2.3) is a red alga with a short stalk and finger-like blades that grows to 15 to 20 inches (37–50 cm) long and 6 inches (15 cm) wide. It grows in deep water near the shore as well as in the intertidal zone. For hundreds of years, this alga has been eaten raw or cooked in Great Britain and Iceland. It has a distinct seaweed taste that may be unpleasant to people who are unaccustomed to it. In some restaurants along the coast of Nova Scotia, it is served like potato chips to accompany soup and sandwiches. In bars it can be found in bowls to increase thirst and encourage patrons to order more drinks.

The nutrient value of dulse is questionable since humans cannot digest many of the complex carbohydrates in seaweeds. It can be a source of trace elements, especially iodine, and perhaps vitamins as well.

Irish Moss (Chondrus crispus, fig. 7.7) is a red alga with a highly branched blade that may be 6 inches (15 cm) high and 4 inches (10 cm) wide. It grows on rocks at the lower edge of the intertidal zone and in deeper water. It contains a substance called carrageenan that is much in demand. Several centuries ago in the British Isles, especially Ireland, Irish moss was collected, dried, and boiled in milk. It was then flavored and allowed to gel as a pudding dessert. Today it is harvested in the maritime provinces of Canada and in New England and shipped to processing plants to extract the carrageenan. This substance is used in everything from dairy products and pharmaceutics to paper making and meat canning, and the uses are growing. In 1969 eastern Canada harvested 43,150 metric tons of Irish moss.

This alga is abundant along the Atlantic coast from New Jersey northward.

Red laver (Porphyra leucostica) is a red alga very

7.7. Irish Moss *(Chondrus crispus)*

similar to sea lettuce (*Ulva lactuca*, fig. 2.1) except for the color. This genus is the most cultivated of all the algae. In Japan they are grown in great plantations in Tokyo Bay. It is sold there in dried sheets or fresh to be cooked with other foods. In making the well-known Japanese dish sushi, it is used to wrap rice and fish or vegetables into a roll. These algae are important sources of food in China, the Philippines, Hawaii, Great Britain, and Ireland. It is reported to have been the primary source of salt in the diet of coastal Indians in North America. Red laver is used in soups, sandwiches, and as a seasoning in the preparation of rice.

This alga grows on rocks and piers along the Atlantic coast from Florida to Newfoundland and the Hudson Bay region.

Winged kelp (*Alaria esculenta*, fig. 7.8) is a brown alga with a short cylindrical stalk. It is olive green to brown in color, grows attached to rocks below the tidal level, and is common where there are very strong surfs and tidal currents. The blade, which may be 1 to 10 feet (.3–3 m) long and more than 10 inches (25 cm) wide has a prominent midrib running its entire length. The stalk has numerous fingerlike small blades attached beneath the main blade.

This alga is one of the main components of a Japanese food called "kombu," which consists of dried sheets compressed into blocks. It is made into soup or cooked with meat or fish. The fingerlike growths are used for food in some regions of Ireland and Scotland. It is not a significant food in North America.

Winged kelp is common along the rocky Atlantic coast from Cape Cod northward.

Sand and Beaches

Beach pea (*Lathyrus maritimus*, fig. 4.7) has clusters of pealike seedpods that develop from the flowers.

The very young, tender seeds, while they are still bright green, can be cooked like garden peas. As they get older, they become dry and tasteless. Although the seeds of this

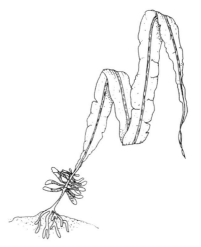

7.8. Winged Kelp (*Alaria esculenta*)

species are safe to eat, the seeds of some other members of this genus are poisonous.

Prickly pear *(Opuntia humifusa,* fig. 3.1) and other members of this genus have been widely eaten by the Indians of the southwest. The pulp of the fruit is flat tasting although sweet and somewhat juicy. The ground seeds can be used as a thickener for soup and stew dishes, and the flat sections of the stems can be peeled and cooked as vegetables. The reddish-purple fruits of prickly pear are usually available in late summer at markets in the east.

Sea-beach sandwort *(Honckenya peploides,* fig. 4.10), when young and tender, may be added to salads, cooked as greens, or made into pickles. Eaten raw it tastes somewhat like cabbage. In Iceland it is used to make a fermented beverage.

Salt Marshes and Salty Soils

Samphire *(Salicornia europaea,* fig. 5.6), perennial glasswort *(S. virginica,* fig. 5.5), and other plants of this genus can be cooked as vegetables, but a change of water is necessary to reduce the saltiness. They also may be pickled in vinegar or chopped and added to salads and coleslaw.

Spearscale *(Atriplex patula,* fig. 5.11) can be cooked as greens or the young tips can be eaten raw in salads. According to several sources they are better than lamb's-quarters, domestic spinach, beets, or chard. Some tribes of Native Americans used the seeds of the plants of this genus to make an edible meal that they mixed with cornmeal.

Southern sea-blite *(Suaeda linearis,* fig. 5.7) can be cooked as greens when young. The water must be changed two or three times to reduce the saltiness. Some tribes of Native Americans added the leaves to other dishes for the salt they provided.

Narrow-leaved cattail *(Typha angustifolia,* fig. 3.6) has a long history of use as a food plant. The heart of the young shoots can be used in salads or as a stir-fry vegetable. The young flower spikes, before the yellow pollen is visible, can be boiled and eaten like corn on the cob. The pollen can be collected in great quantities and used directly or mixed with wheat flour to make bread. The rootstock is very rich in starch that can be extracted and used as flour. Cattails have great potential as a commercial source of food for the future.

8

Naming, Collecting, and Preserving Plants

Plant Names

There are more than 300,000 known species of plants on Earth today. Each species has a name, and some have more than one. When a new species is discovered, the individual who recognized it as new has the honor of giving it a botanical name. Every species thus has a Latinized botanical name. Latin is no longer used by human culture, and so the Latin name will have the same meaning five hundred years from now as it does today. In addition to the botanical name, many plants have one or more common names that usually date to antiquity. These were given by people who were familiar with the plant because it was harmful to humans in some way, was a source of medicine, was useful for food, or had outstanding physical characteristics.

Samphire is a common salt-marsh plant. It is widespread in Europe where it has a name in the language of each country in which it grows. It may even have more than one name in each language. For example, in the United States samphire is also known as slender glasswort, chicken claws, and pigeon foot. Sometimes a particular common name is applied to entirely different plants.

In order for a plant name to be of scientific value it must be the one and only name that universally refers to that plant. If a botanist wishes to publish a paper on research conducted on samphire, the Latin name is used so that the subject of the study will be recognized by other botanists. One of the American common names would mean nothing to botanists in Russia, India, or Germany. This is why research reports always identify plants by their botanical names. The botanical name for samphire is *Salicornia europaea*, and it is spelled the same way in every language of every country in the world. No other plant on the globe has this name.

Common or Folk Names

While every known plant species has a Latinized botanical name, most do not have common or folk names. Possible reasons for this are that many plants are small and escape notice, or they are growing in areas infrequently visited by humans. Among the flowering plants, some closely related species are so similar that an entire cluster of species, called a genus, may be known by a single common name.

Although common names are unsuitable for the identification of plants in research papers, they represent a wealth of information and folklore. Many common names refer to the habitat of the plant. For example, no one will be surprised that seaside goldenrod and beach pea are seashore plants.

Botanical Names

Botanists have been using Latin to name plants for hundreds of years. Before the mid-eighteenth century, these names often consisted of several words and were more like descriptions than names. In 1753 a Swedish botanist named Carl Linnaeus wrote a book entitled *Species Plantarum*, which, freely translated, means "The Species of Plants." In the book he used a two-word system to give names to all the plants in the world known to him. Although it met with resistance from some botanists of the time, this method greatly simplified the naming of plants. This two-word or binomial system is used by botanists today.

There is a well-defined procedure for giving a newly discovered plant species a name. Following the system initiated by Linnaeus, the botanical name consists of a generic name and a specific name. The botanists coming after Linnaeus added a third component to the name: the initials of the botanist who named the plant. All the plants named by Linnaeus have an L. following the specific epithet. For example, the botanical name for samphire is *Salicornia europaea* L. This plant is in the genus *Salicornia*, its specific name is *europaea*, and it was named by Linnaeus. In nontechnical publications, the initials of the botanist who named the plant are often omitted. In writing the botanical name, the genus is always capitalized and the specific epithet should never be written with a capital letter. The word species is both singular and plural as illustrated in the following sentence: "The genus *Salicornia* has at least two species but some genera have only one species."

The botanical name provides both descriptive information about the species and information about its evolutionary relationships. All the species that make up a genus evolved from a common ancestor, thus, a genus is a group of closely related species. Likewise, a group of closely related genera make up a family, and a group of similar families is an order. All of the genera in the aster family evolved from a common ancestral genus and are more similar to one another than they are to the genera of any other family. Most generic names are hundreds of years old and are derived from ancient Latin or Latinized Greek words.

The second word of the botanical name is the specific epithet. It is usually a word that describes some characteristic of the species. For example, *Lathyrus maritimus* is the botanical name for beach pea. *Lathyrus* is an ancient Greek name for bean or pea and *maritimus* means "of the seashore." *Typha* is the genus name for cattails and *angustus* means "narrow." *Typha angustifolia* then, is narrow-leaved cattail. It should be noted that the botanical name for this plant must include both the genus and the specific epithet. Angustifolia is the specific epithet for several plants. Only when it is used with *Typha* does it mean narrow-leaved cattail. This is true for all specific epithets: they are parts of the botanical names only when used with genus names.

Although botanical names are indispensable for professional botanists, for the uninitiated they sometimes seem long and difficult to pronounce. For example, beach-grass is known by the botanical name *Ammophila breviligulata*. Many—perhaps most—botanical names are shorter than this and with experience and a little effort they become much easier to use. One of the pleasures of being familiar with plant names is having the ability to talk with others who have similar interests. For maximum communication with others, at all levels, learning the botanical as well as the common name is recommended.

What Is a Species?

The term "species" has been used frequently in the preceding pages, so a few words of explanation are appropriate. The species is the basic unit of classification. The living representative of a genus, a family, or an order is a species. The genus, family, and order are classification concepts but the species can be seen and touched.

A species is a group of plants that resemble one another more than they do members of other species. The plants in a species interbreed freely but do not interbreed with members of other species. Although these statements are generally accepted as reliable descriptions of a species, they are oversimplifications because sometimes different species can interbreed to produce hybrids. These hybrids are ordinarily sterile and do not produce offspring, but this is not always the result. To complicate matters even more, some plant species can develop seeds without pollination and the subsequent union of male and female sex cells. These seeds germinate and grow into plants that are the exact replicas, or clones, of the parent plant. Every student of botany soon learns that it is difficult to formulate a definition of a species that does not have exceptions. The reader is challenged to explore this topic further in the readings at the end of this book.

Collecting Plants

Humans are collectors of the things that interest them, from bottle caps to vintage cars. It is not surprising, then, that people who are interested in plants should collect plants. It is likely they are individuals who love the outdoors. Collecting plants not only satisfies their collecting desires but also provides exercise and fresh air. It is a pleasurable activity but one that should be pursued with some caution.

Where to Collect

Plant collectors do not have the freedom to collect all the plants they want wherever they see them. Most of the land surface in the United States is owned by someone or some organization. To avoid legal entanglements, permission should be acquired before collecting on private property. Collecting is prohibited on some municipal, county, state, and federal parks, but limited permission can sometimes be obtained if park managers are approached with tact. In all instances, to avoid trespassing, the best practice is to seek permission to collect.

Where and What Not to Collect

Even after permission has been granted to collect in a particular area, the collector is not absolved of all responsibility. Out of consideration for environmental conservation, it is a good practice to follow a few simple rules of conduct. When there are only a few plants of a species growing in an area, it is best to collect where they are more abundant. If there is only one plant of that species growing there, it should *never* be collected. The collecting area should be altered as little as possible by the collector. Collecting the last specimen of a species eliminates the colony from that area.

State conservation departments can provide lists of rare and endangered plants for their states. To avoid extinction of these species, every effort should be made to assure their survival and perpetuation in the natural world. They should not be collected. A suggested alternative to collecting these plants is to collect their seeds and grow your own. Mature seeds can be harvested without damaging the plant and it will be a challenge to try to create environmental conditions under which they will germinate and grow.

Plant collectors should be aware that some plants are poisonous and may cause serious skin rashes on individuals sensitive to them. The most common culprit is poison ivy. About 80 to 85 percent of the population in the United States is allergic to this plant. Even those who are not allergic should avoid contact with such plants because repeated exposure sometimes causes sensitivity to develop. For more details on the identification and the effects of poison ivy, see chapter 7.

Tools for Collecting

Experienced collectors usually have a kit that contains the essential tools. It may be stored in a backpack or in the trunk of a car. The basic items that should be included in the kit are a cutting tool, containers for the specimens, notebook and pencil or pen, identification tags, and a hand lens. These are described in more detail below.

Cutting Tool

Every collector needs a cutting tool of some type such as a penknife or a pair of hand clippers or pruning shears. Any kind of pocket knife with a sharp blade will be satisfactory, but sometimes for woody plants, hand clippers are better.

Plant Containers

The traditional type of container used by professional botanists for plant collecting is called a vasculum. It is usually constructed of a light metal such as aluminum and has an easily opened and closed lid and a shoulder strap. When the vasculum is lined with wet newspaper, it will keep plants from wilting for several days. These containers can be purchased from biological supply houses but are rather expensive. A selection of scientific supply houses are listed at the end of this chapter.

Plastic bags are less expensive, easier to store and transport, and even many professional botanists are finding them more convenient than vascula. For smaller plants, bags with a zipper closure are satisfactory. Larger bags that are closed with a twist-tie can be used for larger specimens. Bags of several different sizes, from sandwich size for small plants to very large ones, should be included in the collecting kit. Experience will teach the best mix of sizes to have available.

To keep collected samples from wilting, a piece of wet newspaper can be placed inside and the bag should be kept out of direct sunlight. Specimens prepared in this manner will remain fresh for two days or more. If the plastic bags or vascula are stored in a refrigerator, the specimens will remain in good condition for up to a week. Under no circumstances should plants be frozen if they are to be pressed or dried. When frozen specimens are thawed, they appear to have been cooked.

Notebook

The importance of a field notebook cannot be overemphasized. A record of each species collected should be made on the spot if possible. A collecting trip may yield several species. If recording the data on these is postponed until the end of the day, details may be forgotten or the collection site of one species may be confused with that of another. The data recorded for each plant collected should include the habitat, such as "dry, sunny hillside," "moist, shady woods," "margin of a cultivated field," "edge of a swamp," and so on. The geographical location should be noted with as much detail as possible. If United States Geological Survey topographical maps are available, rural roads, wetlands, fields, and forests will be identified and the latitude and longitude can be determined. For information on USGS topographical maps and how to acquire them, write to:

United States Geological Survey
Map Distribution
1200 Eads Street
Arlington, VA 22202

Knowing the exact location of the site and the date of collection are important if the collector wishes to return at another season for flowers or fruits. This information can also be helpful to other collectors.

Other items that should be recorded at the time of collection, since they may change with time, are flower color and odor. The number of flower petals should be noted because some may fall off after the plant is placed in a collecting bag.

Identification Tags

Each specimen collected should be identified with a number or letter. Suitable tags can be inexpensively purchased at almost any store that sells office supplies. The field notebook entry should be listed under this number, and a tag should accompany the specimen at all times. It can be attached directly to the plant or placed in a bag with only one specimen in it. The identification number or letter on the tag should be written with a pencil or a pen with ink that does not smudge or smear when moistened.

Hand Lens

An item that may not be essential but can be very useful to the collector is a small hand lens. One that magnifies about ten times is sufficient for most uses. The hand lens is especially useful when examination of flower parts is necessary for the identification of a species.

The Specimen

Herbaceous Plants

When collecting herbaceous plants, the single most important features are the flowers because they are necessary for identification. If the plant is to become part of a collection, it should be in bloom when it is collected. Sometimes identification is easier if both flowers and fruits are available. Some plants bloom over a period of time, so both flowers and fruits can be collected on the same specimen. Usually, though, if fruits are required, it will be necessary to return to the collection site later in the season. Since the flowers are essential for identification, if the plant is unknown to the collector it is useful to collect a few extra. This will allow for the dissection that is often necessary for identification and leave an undamaged specimen for the collection.

The ideal specimen is one that is representative of the species. It should not be the largest or smallest plant in the colony but rather near the size of most plants of that species at that location. The specimen should be in good physical condition with a minimum of insect damage. There should be enough leaves to clearly demonstrate whether they are attached to the stem in pairs (opposite) or singly (alternate).

Sometimes for smaller plants, the entire specimen, including the roots, can be taken. Entire plants should be removed carefully so as not to damage or deface the collecting site. When collectors leave a collection area, it should look exactly the same as before they arrived. For some larger plants, usually only the upper portion of the stem with its leaves and flowers is collected. In addition to a leaf-bearing stem, some plants also have basal leaves that grow directly from the rootstock. Whether they are the same or different from the stem leaves is sometimes an important identifying feature. When plants have basal leaves, a few of these should also be collected.

Ordinarily one specimen of a species is enough for most collectors. If the species is less than abundant, the collector should be guided by good conservation practices and limit the number of samples to one. When a species is plentiful, two complete specimens may be taken in case one becomes damaged. One specimen can sometimes be used for confirmation of identity by sending it to an expert botanist. It is usually not necessary—or good plant conservation practice—to take more than two specimens.

Woody Plants

Many trees and shrubs can be identified by leaves alone but some require fruits or seeds. Nuts are essential in the identification of species of oaks and hickories. The specimen for a tree or a shrub consists of a twig from the end of a branch with enough leaves to clearly show leaf arrangement. Identification almost always requires knowledge of whether the leaves are opposite or alternate. The end of a branch that developed in sunlight is best for this purpose. A twig growing in the shade grows slowly with the distance between alternate leaves so short that they may appear to be in pairs or whorls.

Identifying the Plant

If a plant that has been collected is unknown to the collector, it is easier to identify as a fresh specimen than as a pressed and dried one. It is advisable, then, to identify the plant as soon as possible. Most plant manuals and handbooks include dichotomous keys for the identification of unknown species. Dichotomous keys are based on the assumption that any collection of plants can be divided into two groups by an observable characteristic that is present in one group but not in the other.

When comparing two specimens of the same species, there can be much variation in physical characteristics. For example, two plants grown under different environmental conditions may be different in height and thickness of stems or in the number, shape, and size of leaves. In these same two plants, though, there will be very little variation in flower parts. In most instances, the characteristics are observable with the naked eye but sometimes a hand lens is helpful.

The use of a dichotomous key can best be illustrated by a small group of plants. Consider a collection with the characteristics listed below.

Plant 1: 10 white petals, 5 stamens, several pistils
Plant 2: 5 white petals, 10 stamens, 1 pistil
Plant 3: 3 white petals, 6 stamens, 1 pistil
Plant 4: 6 blue petals, 6 stamens, several pistils
Plant 5: 5 blue petals, 5 stamens, several pistils
Plant 6: 3 blue petals, 3 stamens, 1 pistil
A dichotomous key for these plants is given below.

A. Plants with petals and stamens in numbers divisible by 4 or 5
   B. Petals blue .... Plant 5
   B. Petals white
      C. Pistil 1 .... Plant 2
      C. Pistils more than 1 .... Plant 1
A. Plants with petals and stamens in numbers divisible by 3
   D. Petals white .... Plant 3
   D. Petals blue
      E. Pistil 1 .... Plant 6
      E. Pistils more than 1 .... Plant 4

In this key, the contrasting statements are given in uppercase letters. The user repeatedly places the specimen in one group or another based on its physical characteristics in progressing to the identity of the unknown plant. Two observations are in order for this type of key.

(1) Each of the contrasting statements from which the user must choose are the same number of spaces from the left margin.

(2) The contrasting statements often begin with the same word followed by a word or statement that expresses a different condition, for example, "Petals blue" or "Petals white."

There are ways, other than the one above, that dichotomous keys can be organized, but they all require a series of choices between characteristics in arriving at the identity of an unknown plant. The above key is an oversimplification because it involves a very limited group of plants. The keys in plant manuals are much more complex because they cover a greater number of plants.

Books on plant identification are listed at the end of this book.

Preserving the Collection

A herbarium is a collection of pressed, dried, and mounted plant specimens. The objective of collecting plants for most amateur botanists is to accumulate a personal herbarium. This differs from the professional botanist only to the extent that the latter usually collects for an institution such as a college, university, or botanical garden. In order to be useful, a plant specimen must be properly prepared and include essential collecting information. A method that has been used successfully by botanists for many years is to press the specimen flat, let it dry, then attach it with glue to a sheet of white paper. This procedure is the same for both amateurs and professionals. Mounted in this manner and protected from insects, the specimen will last for hundreds of years.

With the passage of time, natural areas already under stress from human activities will become even further disturbed, if not entirely eliminated, as the human population increases (the United States grows by nearly 2 million people per year). Many plants that are abundant today will without doubt become much less so in the future. The amateur's collection may thus become a valuable documentation of rare, endangered, or extinct plants.

Equipment Needed for Pressing

Collectors will develop individual routines for preparing specimens, but some basic equipment includes a plant press, corrugated cardboard ventilators, blotters or newsprint, mounting paper, labels, and an adhesive. These items, with some alternatives, and their uses are explained below.

The Plant Press

The function of the plant press is to thoroughly flatten the specimen and hold it in place until it dries. This is accomplished with two solid or wooden grid frames held together by two ropes or straps. The plant to be pressed is placed between the two frames, which distribute the pressure evenly. The straps can then be tightened to the desired amount. Plant presses can be purchased at biological supply houses, or they may be constructed inexpensively. Two pieces of quarter-inch plywood or perforated masonite, each 12 by 18 inches, will serve satisfactorily as frames. Two pieces of heavy cord or, preferably, canvas straps with buckles, each about five feet in length, can be used to hold the frames together.

Corrugated Cardboard Ventilators

Cardboard ventilators can be purchased from biological supply houses or they can be made from corrugated cardboard boxes. Ventilators cut from boxes should be 12 by 18 inches with the corrugations parallel to the 12-

inch side. These allow air to pass freely through the plant press for rapid drying of the plant.

Blotters or Newsprint

Blotters or newsprint are in direct contact with the plant and absorb juices that may be squeezed from it as it is pressed. If 12-by-18-inch blotters are not available, pages of newsprint folded in half are approximately 12 by 14 inches and are suitable substitutes. Three or four pages of newsprint folded in half will perform the same function as a blotter.

Mounting Paper

Some collectors may wish to mount their plant collections on the pages of scrapbooks. An advantage of scrapbooks is the great variety in the kinds available and the ease of displaying and viewing the collection. A major disadvantage is that most scrapbook pages are smaller than standard herbarium sheets, which are $11^1/_2$ by $16^1/_2$ inches. This is the size of mounting paper used in all professional herbaria. The personal herbarium of the collector will be of greater value if its specimens are compatible with those of professional herbaria. Mounting paper can be purchased at biological supply houses.

Labels

The sheet on which the specimen is mounted must have a label. Commercial mounting paper can be purchased with the label already printed in the lower-right-hand corner of the sheet. Plain paper is less expensive, and standardized printed labels can be purchased separately or made easily. The label must provide several items of essential information. Obviously the first item should be the name of the plant. The manual that is used to identify the plant will give the botanical and common names, and the label should carry both. The botanical name should be listed first. Sometimes it will be the only name since some species have no common names. In professional herbaria, the initials of the botanist who named the species are included as part of the botanical name.

The label should also give information in as much detail as possible about the location of the collecting site and the habitat from which the plant was collected. Finally, the name of the collector, the specimen number, and the date the collection was made should be listed.

The specimen number deserves a special mention. Some collectors keep a lifetime list of the plants they have identified or collected and number them consecutively from 1 onward. Others prefer to start their numbering anew each year and designate the year of collection as 01–1, 01–2 then 02–1, 02–2 etc. The specimen number, environmental data, and site location will be provided by the field notebook.

All of this information can be recorded on a label about the size of a 3-by-5-inch card. If you make your own, four labels can be typed on a sheet of $8^1/_2$-by-11-inch paper. The following is suggested as a model.

HERBARIUM OF JANE DOE

Botanical Name _____
Common Name _____
Family _____
Locality_____
Habitat _____
Collector _____
Date _____ No. _____

Adhesive

The function of the adhesive is to attach the specimen to the mounting paper. White glue such as Elmer's glue is probably the most convenient for the individual collector. It is readily available from many stores, is very effective, and is used by many professional botanists. Some collectors prefer thin strips of an adhesive linen tape to attach the specimen. This adhesive is available at most office supply stores. Transparent plastic tape is unsatisfactory because it dries and yellows with age.

Pressing the Specimen

Some collectors carry a plant press into the field and press the specimens as soon as they are collected. Others prefer to transport the specimens to a home base where conveniences such as work tables may be available. Regardless of the location, there is a recommended routine for the process as described below.

1. The bottom frame of the press should be placed on the ground or on a table.

2. A corrugated cardboard ventilator is the placed on the frame.

3. A blotter, or in the absence of blotters, several pages of newsprint folded in half, are placed on the frame.

4. The plant specimen to be pressed is placed on one half of a page of folded newsprint. It should be spread carefully so that flowers are unobstructed and there is a minimum overlapping of leaves. One or two leaves should be turned over with the bottom side up, since features of the leaf undersides are sometimes important for identification. If a specimen is too large to fit easily on one-half of a page of newsprint, it can be bent to form a V, or if still larger, bent again to form an N. Then the other half of the newsprint page is folded over the specimen. The name of the plant or its number is written on the outside of the folded newsprint.

5. A blotter, or two or three pages of folded newsprint, are then placed on top of the newsprint containing the plant.

6. Another corrugated cardboard ventilator is placed on the blotter or newsprint.

7. The process can now be repeated for other specimens in the same order: ventilator, blotter, specimen, blotter, ventilator.

8. When all specimens have been so prepared, the top frame of the press is placed on the stack and the straps tightened around each end. Apply as much pressure as possible in tightening the straps. Having someone stand on the press while tightening is helpful.

Drying

The faster the specimen dries, the less likelihood that there will be discoloration of the flowers and leaves. Pressing a plant in a book is not recommended because the specimen dries slowly with practically no air circulation, usually resulting in discoloration of not only the plant but the pages of the book as well. In a plant press, depending on the temperature, humidity, and the size of the specimen, it will dry in five to ten days. After the first twenty-four hours, it can be examined to rearrange flowers or to smooth wrinkles.

If faster drying is desired, the plant press can be positioned over a mild source of heat so that warm air rises through the channels provided by the corrugations of the ventilators. The most convenient source of heat is probably an ordinary light bulb. The press can be placed between two chairs with the bulb at least one foot below the corrugations. Only mild heat is recommended because overheating may cause the specimens to turn brown. The plants will dry in two or three days with this arrangement.

If blotters and ventilators are not available, the plant press can still be useful. On the bottom frame, place a stack of three or four pages of folded newsprint. On top of these place the folded page holding the plant to be pressed. Add another stack of newsprint similar to the first. At this point, other specimens can be added following the same layering process. The top frame of the press can now be applied and the straps tightened. It may take a little longer for the plants to dry by this method, but the porosity of the newsprint will provide enough aeration to prevent discoloration. To hasten the drying process, the newsprint can be changed after the first twenty-four hours.

Mounting

When the plant is removed from the press, it is ready to be attached, or mounted, on a sheet of paper. The traditional method of mounting is to coat a pane of glass with a thin layer of brown glue, lay the dried specimen on the pane to pick up some glue, and then place it on the mounting paper. This is a satisfactory method for a large professional herbarium with many plants to mount, but it is not suitable for the individual collector who may wish to mount only one or two plants at a time.

Using white glue that can be squeezed through a nozzle from a tube or other container, the individual collector can apply dots of glue to several places on the underside of the specimen. After positioning it on the paper, dots of glue can be applied to other points as needed. Sometimes a thin string of glue, when it dries across a leaf or other delicate part, will effectively pin it to the paper. An advantage of using glue is that usually it will last as long as the paper or the plant. A disadvantage is that the plant can never be removed from the sheet.

An alternate method of mounting is to use linen gummed tape. Thin strips of tape can be placed across stems and leaves at critical points to hold the plant on the paper. This method has the advantage of allowing the removal of the specimen from the paper at some future date. A disadvantage is that over a long period of time the tape may dry and lose its adhesiveness.

Attaching the Label

Attaching a label, complete with the name or identifying number of the plant, must be a part of the mounting routine. The most convenient labels are those that are already printed on commercial herbarium paper. However, gummed labels can be purchased. If you make your own, they can be attached to the page with the same glue that was used to attach the specimen.

Protecting and Storing the Collection

There are two major threats to any herbarium: fungi and insects. Preventing contact of the specimen with moisture is the key to controlling fungal growth. If the mounts are dry at all times and stored in an area that has consistently low humidity, the threat of fungal attack is greatly reduced. Insects often cause a greater problem than fungi, however. Even if the plants are completely dry, an infestation may occur. Among the most damaging of the insect pests are several species collectively called dermestid beetles. They are very small beetles that in the larval stages feed on dry plant tissue. The collection should be inspected at least three times a year for indications of fungal or insect damage.

Professional herbaria store their collections in airtight metal cabinets. These can be purchased from biological supply houses, but they are very expensive and probably impractical for the individual collector. Professional herbaria also use large manila folders, called species covers, to hold all the specimens of each species. While these are convenient, they are not essential, and in their place the collector can use folded newsprint pages. Any appropriately sized cabinet or even cardboard boxes will serve as storage facilities. They can be made approximately airtight by splitting large plastic trash bags and tacking or gluing them in as liners. If the collection is mounted in scrapbooks, they can be stored in large plastic bags. It is worth repeating that whatever the storage facility, the storage area should be well ventilated and dry.

There are several types of fumigants that can be used to protect the herbarium from insects. The easiest to acquire is probably paradichlorobenzene (PDB), which can be purchased as either moth crystals or moth balls. If the storage cabinets or boxes have reasonably tight closure, a small cloth bag of crystals or perforated bags of moth balls can be placed in each compartment. Like scrapbooks, smaller collections mounted on individual sheets can be stored in large plastic bags into which crystals or moth balls have been inserted. The chemicals should be renewed about every four months. If the herbarium is stored at home, it should be kept in an area where family members will not be constantly exposed to PDB fumes.

Some botanists have suggested an alternate method of protecting the plant collection from infestation. Placing the mounts in the freezer for twelve to fourteen days seems to be enough to kill insect pests. This has great appeal to many people because it eliminates the use of chemicals. A disadvantage may be that it requires the periodic availability of a considerable amount of freezer space.

Displaying the Collection

There are several ways that plant mounts may be prepared for display. Collectors often give presentations for school groups, scout groups, 4-H clubs, or other organizations for young people. It may be desirable in these presentations to have specimens that can be handled by the audience. Young, eager, and curious hands can do a lot of damage to a dry and very brittle mounted plant. For collections mounted in scrapbooks, the best books are those with individual pages that are removable and have plastic covers. These provide a measure of protection for the plant and are excellent for viewing.

Collectors who do not use scrapbooks may wish to laminate with plastic the mounts of the specimens they will use for a presentation. However, the cost of this option may be prohibitive. An alternative approach is to attach the plant mount to a standard corrugated cardboard ventilator, or other stiff cardboard to prevent bending the specimen, and then wrap it tightly with plastic kitchen wrap. Plants prepared in this way are suitable for hands-on presentations to groups of all ages.

Sometimes special mounting is appropriate for specimens that are bulky or unusually attractive. For mounts that are flat, the whole sheet can be enclosed in a frame called a botanical mount. It consists of a stiff cardboard back with a glass front held together usually by black tape around the edges. The botanical mount may contain a thin layer of cotton to hold the mount in place. For bulky specimens such as those with thick stems, pine cones, or hard fruits, a type of frame known as a Riker mount is available. These are shallow cotton-filled boxes with a pane of glass on one side. The specimen is usually not mounted on a sheet of paper but is held in place by the cotton. Both of these mounts are expensive, but plants mounted in these ways are often so attractive that they can be displayed as wall hangings.

Special Plant Groups

Some collectors may wish to include examples of all major plant groups in their collections. The discussion in the preceding pages has been concerned mainly with methods of collecting and preserving seed plants. These methods are valid for most plants, but some groups require a different type of treatment. The life histories and growth habits of the plant groups listed below are described in chapter 2.

Algae

Algae may be preserved as dry mounts or in preservative liquids. To make a dry mount, hold a piece of mounting paper in the water beneath the alga. Raise the paper very slowly, tilting it to let the water escape but not the alga. The paper containing the specimen can then be allowed to air dry or it can be put into a plant press. If a plant press is used, feel the alga and if it is sticky, cover it with a sheet of waxed paper to keep it from sticking to the newsprint. As it dries, the alga will become attached to the mounting paper, so an adhesive is unnecessary. When the specimen is dry, the paper can be trimmed and attached to a scrapbook page or other mounting sheet and a label added.

The method used to get a specimen of an alga on a mounting paper can also be used for other kinds of aquatic plants, many of which have finely dissected or have very thin leaves. When these are removed from the water, they collapse and become difficult to work with. A piece of mounting paper can be inserted beneath the plant underwater near the surface. The leaves can be arranged as desired and the plant lifted slowly out of the water. After allowing the mounting paper containing the plant to drain on newsprint, it should be placed in a plant press to dry. Drying may take longer than for nonaquatic plants, but the process can be hastened by changing the blotters or newsprint after twenty-four hours. When the aquatic plant is dry, it may be impossible to remove it from the paper on which it was collected. If so, follow the same mounting procedure that was suggested for algae.

Small commercially available collecting bottles or home food containers such as baby food jars can be used for liquid preservatives. A number of preservatives are available from supply houses. One that is especially good for algae is F.A.A., which is a combination of formalin, alcohol, and acetic acid. This is obviously toxic, so it should be handled with care and kept out of the reach of children. A satisfactory preservative that is available from drug stores and supermarkets is rubbing alcohol. Although it is not harmful to the skin, it can be highly toxic if ingested. A problem with using liquid preservatives is the almost unavoidable loss by evaporation. The bottles should be checked monthly and additional preservative added to keep a constant level. The label can be glued to the outside of the container or included within.

Some of the marine algae pose a problem for collectors because of their large size. The microscopic forms and the larger species that will fit on a mounting sheet can be treated in the same manner as described for freshwater algae. But many of the brown algae are much too large for a standard-sized mounting sheet. With these plants, the collector has at least two options. A piece of the alga can be pressed and mounted, or the entire plant can be collected and allowed to dry. In the latter method, the plant can be folded several times when it is still slightly moist, then allowed to continue drying. When it is completely dry, it can be stored in an appropriately labeled plastic bag. As with other dry specimens, moth crystals should be included in the bag. Some of the large brown algae, or kelps, can be soaked in a solution of glycerin, alcohol, and seawater for several days and retain their flexibility during storage. For details of this method, see MacFarlane's *Collecting and Preserving Plants* (1985).

Fungi

Mushrooms and other soft fungi. The structures that are collected are fruiting- or spore-bearing stages. They can be preserved in a liquid preservative or by drying. For storage in liquid, they should be cut near the ground and placed in the preservative immediately. In the absence of a commercial preservative, rubbing alcohol is satisfactory for this purpose.

If soft fungi are to be preserved by drying, they should be dried as quickly as possible because they decay rapidly when moist. Applying some form of mild artificial heat will hasten the process and help prevent the onset of decay. After drying is complete, the specimens can be stored in labeled boxes of the appropriate size. Moth crystals should be included as protection from insect attacks. To avoid decay, it is especially important to keep these specimens dry.

Shelf fungi. Some of the shelf fungi are hard and woody. The collector needs only to break them from the log or stump on which they are growing. They may grow to a very large size, but the collector can collect a specimen of the size that is best suited for the collection. When they are thoroughly dried, they can be stored in boxes or plastic bags with PDB crystals.

Lichens. A good plan for collecting lichens is to use a plastic bag in the field and then transfer them to boxes or envelopes for storage. The specimen should include the small cuplike spore-producing structures of the fungal portion of the lichen. This is very important for identification. Care should be taken in their transport and storage because they are usually dry and brittle and easy to shatter. In the fungi, as in all plant groups, detailed field notes should be made so that a complete label can be attached to each specimen.

Ferns. Ferns can be pressed and mounted in the same way as seed plants. Some special notes on collecting will be useful. It is important that the specimen have spore-bearing structures. In many fern species, these are on the underside of the frond. In those that have dissected leaves, they are on the undersides of the leaflets. When pressing the fern leaf, be sure to turn a few leaflets over so the fruit dots, or sori, can be seen when the leaf is mounted. In other fern species, the spore-bearing structures are on separate stalks. These must be included for complete specimens.

Scientific Supply Houses

A few supply houses are listed below where the equipment described in this chapter may be purchased.

Carolina Biological Supply Company
2700 York Road
Burlington, NC 27215

Central Scientific Company
3300 Cenco Parkway
Franklin Park, IL 60131

Frey Scientific
905 Hickory Lane
P.O. Box 8101
Mansfield, OH 44901

Wards
P.O. Box 92912
Rochester, NY 14692

Sargent-Welch Scientific
911 Commerce Court
Buffalo Grove, IL 60089

9

Activities and Investigations

1. Making Leaf Collections

Many plants, especially woody plants, can be identified by leaf characteristics alone. Leaf collections can thus serve as aids for identification. Waxed leaf collections can also be used for decorations. Alternate methods of making leaf collections are leaf prints and leaf skeletons. Leaf prints give an outline of the main veins of the leaf, and leaf skeletons show the entire lacy network of the internal vascular system.

A. Leaves

A collection of tree leaves is easy to make, and it can be especially helpful in hands-on presentations to youth groups. Fully developed leaves with no insect damage should be selected and pressed for mounting in the manner described for seed plants in chapter 8. They can be mounted one or more leaves per page, as the collector wishes. Leaf collections of woody plants are more useful because these plants can more reliably be identified by leaf characteristics than can herbaceous plants.

An alternate way to make a collection of leaves is to coat them with wax. After they are thoroughly dry, they can be pressed between layers of wax paper with a warm iron. This will apply a thin layer of wax to each side of the leaf. Prepared this way, the leaves can be either mounted on paper or stored in a box or plastic bag and they will last for years. Leaves that are at the peak of autumnal color can be waxed in this manner, and, although the color may fade after a few weeks, they can be used for attractive seasonal decorations.

B. Leaf skeletons

Another way of displaying a leaf collection is by making leaf skeletons. Leaves are made up of a network of tiny veins. When the interconnecting tissue is removed, a beautiful lacy outline of the leaf remains. There is no quick or easy way to make a leaf skeleton, but when successfully done, the result justifies the effort.

There are chemical methods that can be used for skeletonizing leaves, but they require chemicals that may not be readily available to the individual collector. The easiest method is to let nature take its course. To a gallon of water, add two tablespoons of forest soil humus or other rich topsoil, to insure the presence of decay bacteria, and two tablespoons of sugar or heavy syrup to stimulate bacterial growth. Immerse leaves in the solution and let stand for a month in a warm place. Then remove a leaf and wash it with a gentle stream of water to remove the soft tissue. If a month is not enough time for complete leaf decay, let the solution stand for another two or three weeks. The leaf skeletons can be mounted between glass plates for projection on a screen, or they can be attractively displayed in a scrapbook or on mounting sheets.

C. Leaf prints

Another way of making a leaf collection is to make a collection of leaf prints. This involves the use of printer's ink, which can be acquired at print shops or office supply stores that sell ink for rubber stamp pads. Several methods have been recommended, but one that is simple and produces good results is described below.

With a small, soft brush, such as a pastry brush, apply printer's ink to a piece of Masonite or other hard, smooth surface. Place the underside of the leaf on the inked surface. Cover it with a sheet of newsprint and roll it very gently with only the weight of a rolling pin or a large drinking glass. The objective is to bring all the veins on the underside of the leaf into contact with the ink. Remove the leaf from the inked surface and carefully place the inked side on a sheet of white paper. Cover it with newsprint and again roll it gently, making sure it does not move. The quality of the print will be determined by the amount of ink picked up by the underside of the leaf. With practice, this can be controlled by the amount of pressure applied when the leaf is on the inked surface.

2. Seeds

A. How Do Seeds Travel?

The seed is a remarkable structure. It is a very compact package that encloses an embryonic plant, representing the next generation, and stored food to support the new plant until it can make its own. This is wrapped in a tight protective coat that often has structures that serve as mechanisms for transporting the package to a new area. Such efficiency took millions of years to evolve, and its success is confirmed every time we look upon a landscape covered with plants.

Most plants have evolved adaptations for seed dispersal. Make a collection of seashore plant seeds and try to determine the mode of dispersal. Review chapter 3 and keep in mind the following attributes.

1. The presence of colorful, soft, fleshy fruits is an adaptation that attracts animal predators, which eat the fruits and transport the seeds in their intestines.

2. Hooks, barbs, burs, and sticky seeds are adaptations for dispersal that allow the seeds to hitchhike rides on the fur or feathers of animals.

3. Flattened areas, or wings, and parachutes of hair on seeds are adaptations for dispersal by air currents.

4. Highly branched, bushy plants that break at the base and roll have adapted for seed dispersal by wind.

5. Some seeds have no observable structures for dispersal.

(a) If they are very small, they may be dispersed by wind or in mud on the feet of birds.

(b) If they are large, they may be used for food by animals that transport and store them.

B. Seeds and Roots

Plants absorb water from the soil through elongated single-celled structures called root hairs. These are located at the tip of a growing root. The only place water can enter the plant is through the surfaces of the root hairs. Since water does not always move through the soil toward the root, the root must continuously grow into new untapped areas to keep the plant supplied with water. Relative to the tip of the root, the root hair zone is always the same size and in the same location. New root hairs form constantly near the root tip while those farthest from the tip wither and die as the root grows through the soil.

The necessity for continual growth into new water sources creates an amazing network of roots. The root system of a single plant of rye *(Secale cereale)* was studied by a botanist who found it had a total of 618 kilometers (386 miles) of root length. The root hairs of the rye plant were determined to have a total length of 10,628 kilometers (6,638 miles) and a surface area of 400 square meters (4,444 square feet).

To observe root hairs, line a transparent drinking glass with several layers of thoroughly moistened paper towels. Place radish seeds between the paper and glass. Fill the glass with sand, peat, or tightly packed shredded paper. Add water to this filler daily to keep it moistened. Within a few days root hairs will be visible. Notice that the shortest ones are nearest the tip, the longest ones farther back. Mark the location of the root hair zone on the glass. Allow the root to grow for a few more days. Does the position of the root hair zone change?

3. Life History Investigations

As one becomes more interested in plants of the seashore, it is but a short step to life history investigations. These can be fascinating field activities that require keen observational skills and good record keeping. In addition, they can yield original information because life history studies have not been conducted on every species of plant. In an investigation of this type, observations should be made on every aspect of the life history of the species. The following are suggestions for the events and characteristics to be recorded for flowering plants. This list is not intended to be all-inclusive. As you become familiar with a species, you may wish to add other observations.

A. Seeds

1. Earliest date of germination
2. Number of seed leaves (monocot or dicot)

3. Type of fruit (fleshy or dry)
4. Number of seeds per fruit
5. Number of seeds per plant
6. Size and weight of seeds
7. Seed modifications for dispersal
8. Date and method of seed dispersal
9. Period of chilling needed before germination

B. Stems and Leaves

1. Rate of stem growth in centimeters per week
2. Number, location, and arrangement of branches
3. Date at which the stem stops growing in height
4. Height of stem at maturity
5. Date at which the stem achieves winter conditions
6. Description of leaves (basal, stem, size, sessile, color, etc.)
7. Arrangement of leaves (alternate, opposite, whorled)
8. Type of leaves (simple, compound, lobed, entire, pinnate, palmate)
9. Insects that feed on the plant
10. Date of leaf fall or behavior as winter approaches
11. The manner in which the plant survives the winter
12. Nature of aboveground parts of the plant in winter

C. Flowers

1. Date of appearance of first flower
2. Date of maximum blooming
3. Number and distribution of flowers
4. Number of flower parts
5. Agents of pollination
6. Date of pollen dispersal
7. Life span for each flower

D. Other observations

1. Life span of plant (annual, biennial, perennial)
2. Characteristics of root system

3. Habitat of plant (beaches, sand dunes, salt marshes, intertidal zones, etc.)
4. Type of vegetative reproduction, if any
5. Outstanding features of the plant
6. Stages of life cycle when it may be edible, medicinal, poisonous, etc.

4. Photographing Plants

A camera can be a great asset for the plant naturalist. Almost any kind of camera will suffice but one that allows close-up focusing is recommended. It also can be useful to have a camera with film that can be processed into slides for projection on a screen. Pictures of the collecting sites can also be taken. These provide an added measure of authenticity when attached to the mounted specimens. In addition, a picture of a plant as it grows in the wild is often helpful in identification.

Some individuals confine their collecting to what can be captured on film. This usually results in large numbers of color slides of wildflowers. As the slides accumulate, a system of organization for storage becomes a necessity. They may be organized by habitat, such as plants of beaches, sand dunes, or salt marshes; or by geography, such as plants of Massachusetts or South Carolina. As familiarity with botanical classification increases, the collector may want to organize the slide collection by plant family, such as plants of the lily family or aster family. A collection of plant slides, organized in any way the collector chooses, is rewarding for private or public showings. Specially constructed boxes for slide storage can be purchased from supply houses.

The camera can also be used for time-lapse photography. Taking daily photographs of a germinating seed or hourly photographs of a flower as it opens can yield spectacular results. Setting a camera tripod in exactly the same location for color photographs of collecting sites in each ecological season will provide valuable life history information.

5. Drying Plants Without a Plant Press

Dried flowers are often used in wreaths, swags, and other home decorations. These are obviously not dried in plant presses; there are alternate ways of drying when the objective is a three-dimensional rather than flat specimen. The simplest method is to collect the plants at their flowering peak, or when they are in fruit, tie them in bundles, and hang them upside down in a dry, dust-free, protected place. Plants with many small flowers clustered in dense heads lend themselves to this kind of drying. A few examples of this type of plant are salt-marsh fleabane *(Pluchea odorata)*, sea lavender *(Limonium caroliniana)*, seaside-goldenrod *(Solidago sempervirens)*, horse-mint *(Monarda punctata)*, and horseweed *(Conyza canadensis)*.

A method that has long been used to dry flowers is to bury them in sand. To use this method, cover the bottom of a container with one or two inches of clean, sifted beach sand. Place the flowers on this layer, stems up, and very carefully cover them, making sure the sand is between and around each petal and delicate part, to hold them in their natural positions. Cover the flowers with one or two inches of sand and store the container in a dry place for about two weeks. Then very carefully pour off some of the sand to see if the petals are stiff and dry. If they are not, more drying time will be necessary.

Sand is not a drying agent. It serves as a frame to hold the buried flowers while they dry naturally. Other substances that have been suggested to serve this function are corn meal, diatomaceous earth, powdered pumice, and even dry cereals such as cream of wheat. Very often these substances are mixed with an active dehydrating agent such as borax. Borax and sand, borax and cornmeal, powdered pumice and corn meal, and pure uniodized salt have all been recommended as mediums for drying flowers.

A commercial dehydrating agent called silica gel is widely used for more rapid drying. It has a sandlike consistency and usually contains crystals of cobalt chloride. These are indicator crystals that are blue when the substance is dry but turn pink when the silica gel has absorbed all the water it can hold. An advantage of using this agent is that it can be dried in a regular or microwave oven and used again and again. When the crystals return to their blue color, the drying agent is ready for reuse.

To dry flowers with silica gel, the method is the same as that suggested for sand except it should be in a container with a lid. The container should be closed during the drying process, which may take two to seven days, depending on the number and size of the flowers being dried. For microwave drying, the time is much less. Silica gel should be handled with care and kept out of the reach of children because the dust can cause irritation to respiratory tissues. Silica gel is available from drug stores and craft stores, and the container usually has detailed directions for its use.

9.1. Leaves and stems of (1) grasses, (2) sedges, and (3) rushes

6. Field Identification of Grasses, Sedges, and Rushes

Members of these three families are very commonly found on seashores throughout the world. They are large families and identification to the species level in each is often based on technical traits that are daunting not only to beginners but sometimes to experts as well. Even identification to the family level can be challenging to the inexperienced because the plants in these families all look like grass in general appearance. The following suggestions are offered as fairly simple ways to distinguish between them in the field. This method of identification is not foolproof, and it will not work for every species of each family. If identification to the genus or species level is desired, the reader should consult the books listed in the bibliography by Gleason and Cronquist, Fasset, and Muenscher.

The grass family *(Poaceae,* fig. 9.1) is one of the largest plant families with at least four thousand species worldwide. It is of great economic importance because it includes the cereal grasses such as wheat, oats, rice, and corn, which provide food for most of the world's population. Characteristics of the grass family that can be observed in the field are as follows.

A. The areas of stems where leaves are attached are called nodes. In grasses the nodes are slightly enlarged.

B. Grass stems are usually circular in cross-section.

C. Grass stem internodes are hollow.

D. The base of the grass leaf forms a sheath around the stem that is open on one side.

The sedge family *(Cyperaceae,* fig. 9.1) is less abundant than the grasses but still very common in wetlands, especially in cooler portions of both the southern and northern hemispheres. The tissue in the center of the stems is called the pith. All sedges have it, and the pith of one species, papyrus plant *(Cyperus papyrus),* was used by the Egyptians for making paper. Characteristics of the sedge family that can be recognized in the field are as follows.

A. Nodes are not enlarged.

B. The center of the stem is solid, not hollow.

C. The stem is triangular in cross-section.

D. The base of the leaf forms a sheath that is closed around the stem.

The rush family *(Juncaceae,* fig. 9.1) includes only two genera in North America, the rushes *(Juncus* spp.) and the wood rushes *(Luzula* spp.). The wood rushes are fewer in number and are found mostly in open spaces or woodlands. *Juncus* is the largest and most common genus and the one most likely to be seen in eastern wetlands. One of the differences between these genera is that the sheath around the stem is closed in the wood rushes and open in the rushes. Characteristics of the rush family, specifically the genus *Juncus,* that can be recognized in the field are as follows.

A. Nodes are not enlarged.

B. The center of the stem is solid, not hollow.

C. The stem is circular in cross-section.

D. The base of the leaf forms a sheath around the stem that is open on one side.

Glossary

Bibliography and Further Reading

Index

Glossary

adaptation: A characteristic of an organism that contributes to its survival under the conditions of the environment.
aeration: The process of adding air.
alternate leaf: A leaf arrangement in which there is one leaf at each node.
annual plant: A plant that completes its life cycle in one year and then dies.
anther: The part of the stamen that produces pollen.
axil: The angle between the leaf and the stem.
biennial plant: A plant that lives for two years, producing flowers and seeds in the second year.
biomass: The total amount of organic matter produced by a plant or in a given area.
blade: The flat expanded portion of a leaf.
calyx: The sepals collectively.
canopy: The continuous cover over the forest floor formed by the crowns of the tallest trees.
climax vegetation: The final stages in ecological succession, composed of species that can reproduce themselves rather than being replaced by other species.
clone: A plant that is genetically identical to its parent plant.
compound leaf: A leaf in which the blade is divided into leaflets.
corolla: The petals of a flower.
cuticle: A waxy covering on all the aboveground parts of a plant.
deciduous plants: Plants that lose their leaves at the end of the growing season as opposed to evergreen plants.
diploid: A condition in which cells contain two full sets of chromosomes, one set from the egg and one from the sperm. Zygotes and sporophytes normally are diploid.

disk flower: A tiny flower on the central disk in the flower head of the aster family, as distinct from ray flowers.

dissected leaf: A leaf that is divided into many narrow segments as in some ferns.

ecological succession: The natural replacement of one plant community by another, culminating in climax vegetation.

ecosystem: A community of living things and all the physical factors that make up an environment.

fertile: Capable of sexual reproduction.

fertilization: The union of two haploid gametes, resulting in a diploid zygote.

frond: The leaf of a fern.

gamete: A haploid sex cell such as an egg or a sperm.

gametophyte: A haploid gamete-producing structure or plant.

genus (plural **genera**): A group of closely related plants with a common ancestor. The first word of the two-word scientific name.

germinate: To resume growth, for example, as a seed or a dormant spore or zygote.

girdle: To remove a ring of bark around the trunk of a tree.

groundwater: The water in the ground in the saturated zone or below the water table.

habitat: The environment of an organism or a community.

haploid: Having only one set of chromosomes as in gametes, spores, and gametophytes.

herb: A nonwoody plant that dies back to the ground at the end of the growing season; plants used in medicine or for seasoning.

herbaceous: Having the characteristics of a herb; green, having the texture of leaves, with nonwoody tissue.

herbalist: One who collects, sells, or prescribes medicinal herbs.

holdfast: The enlarged lower end of a macroscopic marine alga that attaches it to a rock or other solid object.

hydrophyte: A plant that grows in a wet environment where it is partially or completely submerged.

internode: The portion of the stem where no leaves are attached; the space between nodes.

intertidal zone: That part of the coast exposed by the water between low tide and high tide.

leaflet: One of the divisions that make up a compound leaf.

mesophyte: A plant that grows in environmental conditions that are intermediate with regard to moisture; between hydrophytic and xerophytic.

micelle: A very tiny soil particle.

morbid: Unnatural; not sound or healthy; diseased.

nectar: A sweet fluid produced by flowers to attract pollinators.

node: The location on a stem where one or more leaves are attached.
opposite leaves: A leaf arrangement with two leaves per node; leaves attached in pairs.
organic matter: Living or once-living tissue; carbon compounds formed by living things.
ovary: The enlarge basal portion of the pistil that contains the ovules and develops into the fruit.
ovule: An embryonic structure inside the ovary that will become a seed.
palmate: In compound leaves, an arrangement in which leaflets are attached at one point and radiate outward as the fingers from the palm of the hand.
perennial plant: A plant that lives for more than two years; not annual or biennial.
petals: The colorful segments of flowers that attract pollinators.
petiole: The stalk of a leaf.
phyte: A suffix that means plant, usually preceded by a descriptive prefix such as hydrophyte, xerophyte, gametophyte.
pinnate: A leaf form in compound leaves in which the leaflets are attached to each side of a central midrib.
pioneer species: The first plants to colonize bare soil or rock.
pistil: The female reproductive part of a flower; the seed-bearing part, consisting of a style, stigma, and ovary.
plant community: All the plant species growing in an area.
pollination: The transfer of pollen from the anther to the stigma.
potherb: A herbaceous plant that is edible when cooked, including the leaves and sometimes the stem.
radial: Spreading outward from a central point.
ray flower: A marginal strap-shaped flower of the aster family.
rhizome: A creeping, horizontal underground stem.
salinity: The degree of saltiness.
sepals: The outermost parts of the flower, usually green and leaflike, which cover the outer parts of the bud.
shrub: A woody perennial not as large as a tree, usually with more than one stem.
simple leaf: A leaf that has a blade not divided into leaflets.
sp.: An abbreviation that follows the name of a genus and indicates a single unnamed or unknown species; *Acer sp.*
species: A group of organisms that can interbreed with one another but not with members of other species.
sporophyte: A diploid plant that produces haploid spores in plants that have alternation of generations.
spp.: An abbreviation that follows the name of a genus and indicates more than one unnamed or unknown species.

stamen: The male or pollen-producing structure of a flower consisting of an anther and a filament.

stigma: The part of the pistil that receives pollen and where the pollen germinates.

style: Usually a slender stalk with the stigma at one end and attached to the ovary at the other.

subspecies: A geographical race of a species.

substrate: Foundation material that makes up a given area of the earth. For example, a bog has an organic substrate.

succession: See **ecological succession**.

succulent: Thick, juicy, fleshy, as in the leaves and stems of plants adapted for dry environments.

summergreen: Description of the eastern deciduous forests that are green in the summer only, as opposed to evergreen.

terrestrial: A land plant as opposed to aquatic.

thallus: A plant body that is not modified into root, stem, and leaf, as in some of the liverworts.

transpiration: The loss of water by evaporation from the surface of plants.

turion: A bulblike structure that serves as a winter bud.

understory trees: Trees that grow beneath the canopy of a forest but do not become part of the canopy.

vegetation: The sum of all the plants.

viable: Alive and capable of growth, for example, a seed.

water table: The top surface of the groundwater.

whorled leaves: An arrangement of leaves with three or more attached at a node.

windfall: Trees blown down by the wind.

wort: A suffix that means plant.

xerophyte: A plant adapted to live under dry conditions.

zygote: A diploid cell formed by the union of two haploid gametes.

Bibliography and Further Reading

Abramaovitz, Janet N. 1996. *Imperiled Waters, Impoverished Future: The Decline of Freshwater Ecosystems.* Worldwatch Paper 128, Washington, D.C.: Worldwatch Institute.

Anderson, Frank J. 1997. *An Illustrated History of the Herbals.* New York: Columbia University Press.

Arms, Karen. 1990. *Environmental Science.* Chicago: Saunders College Publishing.

Bailey, Liberty Hyde. 1933. *How Plants Get Their Names.* New York: Macmillan Co.

Barbour, M. G., J. H. Burk, and W. D. Pitts. 1980. *Terrestrial Plant Ecology.* Menlo Park, Calif.: Benjamin/Cummings Publishing Co.

Bell, C. Richie, and B. J. Taylor. 1982. *Florida Wild Flowers.* Chapel Hill, N.C.: Laurel Hill Press.

Berlin, Brent. 1973. "Folk Systematics in Relation to Biological Classification and Nomenclature." *Annual Review of Ecology and Systematics,* vol. 4. Palo Alto, Calif: Annual Reviews, Inc.

Bold, Harold C., C. J. Alexopoulos, and T. Delevoryas. 1987. *Morphology of Plants and Fungi.* New York: Harper and Row.

Brayshaw, T. Christopher. 1996. *Plant Collecting for the Amateur.* Victoria, British Columbia: Royal British Columbia Museum.

Brown, Lester, et. al. 1990. *State of the World 1990.* New York: W. W. Norton and Co.

Buchholz, Rogene A. 1998. *Principles of Environmental Management.* 2d ed. Upper Saddle River, N.J.: Prentice Hall.

Campbell, F. T. 1980. "Conserving Our Wild Plant Heritage." *Environment* 22 (9): 14–20.

Carlson, Eric, D. Cusick, and C. Taylor. 1992. *The Complete Book of Nature Crafts.* Emmaus, Pa.:Rodale Press.

Carson, Rachel L. 1951. *The Sea Around Us*. New York: Oxford University Press.
Chiras, Daniel D. 1988. *Environmental Science: A Framework for Decision Making*. Menlo Park, Calif.: Benjamin/Cummings Publishing Co., Inc.
Chiras, Daniel D. 1992. *Lessons From Nature*. Washington, D.C.: Island Press.
Christensen, N. L., R. B. Burchell, A. Liggett, and E. L. Simms. 1981. "The Structure and Development of Pocosin Vegetation." In *Pocosin Wetlands*, edited by C. J. Richardson. Stroudsburg, Pa.: Hutchinson Ross Publishing Co.
Cobb, B. 1977. *A Field Guide to the Ferns and Their Families of Northeastern and Central North America*. Boston: Houghton Mifflin Co.
Coffey, Timothy. 1993. *The History and Folklore of North American Wildflowers*. New York: Facts on File, Inc.
Cope, Edward A. 1992. *Pinophyta (Gymnosperms) of New York State*. New York State Museum Bulletin No. 483. Albany, N.Y.: State Education Department.
Courtenay, B., and H. H. Burdsall, Jr. 1984. *A Field Guide to Mushrooms and Their Relatives*. New York: Van Nostrand Reinhold.
Cox, Donald D. 1985. *Common Flowering Plants of the Northeast*. Albany, N.Y.: SUNY Press.
———. 1996. *Seaway Trail Wildguide to Natural History*. Sackets Harbor, N.Y.: Seaway Trail Foundation.
Crawley, M. J., ed. 1986. *Plant Ecology*. Boston: Blackwell Scientific Publications.
Croom, Edward M. 1983. "Documenting and Evaluating Herbal Remedies." *Economic Botany* 37 (1): 13–27.
Cutter, Susan L., H. L. Renwick, and W. H. Renwick. 1985. *Exploitation, Conservation, Preservation*. Totowa, N.J.: Rowman and Allanheld Publishers.
Daiber, Franklin C. 1986. *Conservation of Tidal Marshes*. New York: Van Nostrand Reinhold Co.
Daws, Clinton J. 1981. *Marine Botany*. New York: John Wiley and Sons.
Duxbury, Alyn C., and A. B. Duxbury. 1991. *An Introduction to the World's Oceans*. 3rd ed. Dubuque, Ia.: William C. Brown Publishers.
Fahn, Abraham, and E. Werker. 1972. "Anatomical Mechanisms of Seed Dispersal." In *Seed Biology*, vol. 1, edited by T. T. Kozlowski, 151–221. New York: Academic Press.
Fasset, N. C. 1976. *Manual of Aquatic Plants*, revised by E. C. Ogden. Madison: University of Wisconsin Press.
Fay, Peter. 1983. *The Blue Greens*. Baltimore, Md.: Edward Arnold.
Fernald, Merritt L. 1970. *Gray's Manual of Botany*. 8th ed. New York: D. Van Nostrand Co.
Garrels, Robert M., and F. T. Mackenzie. 1971. *Evolution of Sedimentary Rocks*. New York: W. W. Norton and Co., Inc.

Gibbons, Euell. 1966. *Stalking the Heathful Herbs*. Brattleboro, Vt.: Alan C. Hood and Co., Inc.
———. 1962. *Stalking the Wild Asparagus*. New York: David McKay Co., Inc.
Gibbons, Euell, and G. Tucker. 1979. *Euell Gibbons' Handbook of Edible Wild Plants*. Virginia Beach: Unilaw Library Press.
Given, David R. 1994. *Principles and Practice of Plant Conservation*. Portland, Ore.: Timber Press.
Gleason, Henry A., and A. Cronquist. 1991. *Manual of Vascular Plants of Northeastern United States*. Bronx, N.Y.: New York Botanical Garden.
Hale, M. E. 1979. *How to Know Lichens*. 2d ed. Dubuque, Ia.: William C. Brown, Co.
Hardin, Garrett. 1990. "The Tragedy of the Commons." *Science* 162: 1243–48.
Hardin, James W., and J. M. Arena. 1974. *Human Poisoning from Native and Cultivated Plants*. Durham, N.C.: Duke University Press.
Harper, J. L., P. H. Lovell, and K. G. Moore. 1970. "The Shapes and Sizes of Seeds." In *Annual Review of Ecology and Systematics*, vol. 1, edited by R. F. Johnston, P. W. Frank, and C. D. Michener, 327–56. Palo Alto, Calif: Annual Reviews.
Hillison, C. J. 1977. *Seaweeds*. University Park, Pa.: Pennsylvania State University Press.
Hitchcock, S. T. 1980. *Gather Ye Wild Things*. New York: Harper and Row.
Howe, Henry F., and J. Smallwood. 1982. "Ecology of Seed Dispersal." In *Annual Review of Ecology and Systematics*, vol. 13, edited by R. F. Johnston, P. W. Frank, and C. D. Michener, 201–28. Palo Alto, Calif: Annual Reviews, Inc.
Ingmanson, Dale E., and W. J. Wallace. 1989. *Oceanography: An Introduction*. Belmont, Calif.: Wadsworth Publishing Co.
Jensen, William A., and F. B. Salisbury. 1972. *Botany: An Ecological Approach*. Belmont, Calif.: Wadsworth Publishing Co., Inc.
Joosten, Titia. 1988. *Flower Drying with a Microwave: Techniques and Projects*. New York: Sterling Publishing Co., Inc.
Kaufman, Peter B., T. F. Carson, P. Dayanandan, M. L. Evans, J. B. Fisher, C. Parks, and J. R. Wells. 1991. *Plants: Their Biology and Importance*. 2d ed. Philadelphia, Pa.: Harper and Row.
Ketchledge, E. H. 1970. *Plant Collecting: A Guide to the Preparation of a Plant Collection*. Syracuse, N.Y.: State University of New York College of Environmental Science and Forestry.
Kinghorn, A. Douglas. 1979. *Toxic Plants*. New York: Columbia University Press.
Kingsbury, John M. 1964. *Poisonous Plants of the United States and Canada*. Englewood Cliffs, N.J.: Prentice-Hall, Inc.
———. 1970. *The Rocky Shore*. Old Greenwich, Conn.: Chatham Press, Inc.

Koopowitz, Harold, and Hilary Kaye. 1983. *Plant Extinction: A Global Crisis.* Washington, D.C.: Stone Wall Press, Inc.

Kowalchik, Claire, and W. H. Hylton, eds. 1987. *Rodale's Illustrated Encyclopedia of Herbs.* Emmaus, Pa.: Rodale Press.

Kraus, E. Jean. 1988. *A Guide to Ocean Dune Plants Common to North Carolina.* Chapel Hill: University of North Carolina Press.

Krochmal, Connie, and Arnold Krochmal. 1973. *A Guide to the Medicinal Plants of the United States.* New York: Quadrangle: New York Times Book Co.

Kunzig, Robert. 1999. *The Restless Sea.* New York: W. W. Norton and Co.

Lampe, Kenneth F., and M. A. McCann. 1985. *AMA Handbook of Poisonous and Injurious Plants.* Chicago: American Medical Association.

Lincoff, G. H. 1981. *The Audubon Society Field Guide to North America Mushrooms.* New York: Alfred A. Knopf.

Litovitz, Toby L., L. R. Clark, and R. A. Solway. 1993. *Annual Report of the American Association of Poison Control Centers.* Washington, D.C.: American Association of Poison Control Centers.

Lyman, Francesca, I. Mintzer, K. Courrier, and J. Mackenzie. 1990. *The Greenhouse Trap.* Boston: Beacon Press.

MacFarlane, R. B. 1985. *Collecting and Preserving Plants for Science and Pleasure.* New York: Arco Publishing, Inc.

Mauseth, James D. 1991. *Botany: An Introduction to Plant Biology.* Fort Worth, Tex.: Saunders College Publishing.

McGinn, Ann Platt. 1999. *Safeguarding the Health of the Oceans.* Washington, D.C.: Worldwatch Institute.

Meeuse, B. J. D. 1961. *The Story of Pollination.* New York: Ronald Press Co.

Miller, G. Tyler., Jr. 1992. *Living in the Environment.* Belmont, Calif.: Wadsworth Publishing Co.

Millspaugh, Charles F. 1974. *American Medical Plants.* New York: Dover Publications, Inc.

Mitsch, William J., and J. G. Grosselink. 1993. *Wetlands.* 2d ed. New York: Van Nostrand Reinhold Co.

Muenscher, Walter Conrad. 1972. *Aquatic Plants of the United States.* Ithaca, N.Y.: Cornell University Press.

National Park Service. 1980. *Assateague Island.* Washington, D.C.: U.S. Department of the Interior.

Niering, William A. 1985. *Wetlands.* New York: Alfred A. Knopf, Inc.

Niering, William A. 1991. *Wetlands of North America.* Charlottesville, Va.: Thomasson-Grant, Inc.

Niering, William, and N. Olmstead. 1979. *Audubon Society Field Guide to North American Wildflowers (Eastern Region).* New York: Alfred A. Knopf Co.

Niering, William A., and R. S. Scott. 1980. "Vegetation Patterns and Processes in the New England Salt Marshes." *Bioscience* 30 (5): 301–7.

Oosting, H. J. 1954. "Ecological Process and Vegetation of the Maritime Strand in the Southeastern United States." *Botanical Review* 20 (4): 226–62.

Owen, Oliver S. 1985. *Natural Resource Conservation: An Ecological Approach.* New York: Macmillan Publishing Co.

Peterson, Lee. 1977. *A Field Guide to Edible Wild Plants.* Boston: Houghton Mifflin Co.

Petry, Loren C. 1963. *A Beachcombers Botany.* Chatham, Mass.: Chatham Conservation Foundation, Inc.

Pritchard, Hayden N., and P. T. Bradt. 1984. *Biology of Nonvascular Plants.* St. Louis: Times Mirror/Mosby College Publishing.

Saunders, C. F. 1948. *Edible and Useful Wild Plants of the United States and Canada.* New York: Dover Publications, Inc.

Simpson, Beryl Brintnall, and M. Conner-Ogorzaly. 1986. *Economic Botany: Plants in Our World.* New York: McGraw-Hill Co.

Small, John Kunkel. 1933. *Manual of Southeastern Flora.* New York: Hafner Publishing Co.

Spitsbergen, Judith M. 1980. *Seacoast Life.* Chapel Hill: University of North Carolina Press.

Stebbins, G. Ledyard. 1971. "Adaptive Radiation of Reproductive Characteristics in Angiosperms II: Seeds and Seedlings." In *Annual Review of Ecology and Systematics*, vol. 2, edited by R. F. Johnston, P. W. Frank, and C. D. Michener, 237–60. Palo Alto, Calif: Annual Reviews, Inc.

Steffanides, George F. 1963. *The Scientist's Thesaurus.* Boston: Best Printers, Inc.

Teal, John and Mildred. 1969. *Life and Death of the Salt Marsh.* New York: Ballantine Books.

Thurman, Harold V. 1990. *Essentials of Oceanography.* Columbus, Ohio: Merrill Publishing Co.

Tiner, Ralph W., Jr. 1987. *A Field Guide to Coastal Wetland Plants of the Northeastern United States.* Amherst: University of Massachusetts Press.

Turner, Nancy J., and A. F. Szczawinski. 1991. *Common Poisonous Plants and Mushrooms of North America.* Portland, Ore.: Timber Press.

Van der Pijl, L. 1972. *Principles of Dispersal in Higher Plants.* 2d ed. New York: Springer-Verlag.

Vickery, Margaret L. 1984, *Ecology of Tropical Plants.* New York: John Wiley and Sons.

Waisel, Yoav. 1972. *Biology of Halophytes.* New York: Academic Press.

Weber, Peter. 1993. *Abandoned Seas: Reversing the Decline of the Oceans.* Washington, D.C.: Worldwatch Institute.

Index

Italic page number denotes illustration.

Abscisic acid, 76
Acid rain, 14
Afalinus, salt-marsh, 52, 69, 87
Agalinus maritima, 52, 69, 70, 87
Alaria esculenta, 107
Algae: blue-green, 27; brown, 29; green, 28; red, 30
Alkaloid, 93
Ambrosia artemisiifolia, 95
Ammophila arenaria, 55
Ammophila breviligulata, 55, 111
Angiosperms, *41*
Arrowleaf morning glory, 63
Artemisia campestris, 53, 82
Artemisia stelleriana, 50, 59, *60*, 78
Asclepias lanceolata, 46, *47*
Ascocarp, 31
Aster: annual salt-marsh, 72, 89; perennial salt-marsh, *42*, *43*, 71, 89
Aster subulatus, 72, 89
Aster tenuifolius, *42*, *43*, 71, 89
Atlantic coastal plain, 54
Atlantic white cedar, *39*, 53
Atriplex arenaria, *84*
Atriplex patula, 49, 71, 86, 104, 108

Avicennia germinans, 25
Azolla caroliniana, 36, *37*

Baccharis halimifolia, 61, 71
Batis maritima, 73
Bay: Chesapeake, 20, Delaware, 20
Bayberry, 36, 50, *63*
Beach-grass, 55, 111
Beach pea, 50, *59*, 79, 109, 111
Beach-plum, 36, *51*
Beach wormwood, 59, *60*, 78
Bear oak, 64, 94
Big Cypress Swamp, 24
Biological pump, 13
Biomass, 66
Black mangrove, *25*
Blue-eyed grass, eastern, *78*
Blue-green algae, 27
Boletus, rough-stemmed, 33
Boletus scaber, 33
Borrichia frutescens, 72, *84*
Botrychium dissectum, 38
Bracken fern, 38, *98*
Bracket fungi, 31, *33*

Brass buttons, 73
British soldiers, 36
Broad-leaved kelp, *19*, 75
Broom crowberry, 77
Brown algae, 29

Caffeine, 94
Cakile edentula, *58*, 80
Cakile harperi, 58
Cakile maritima, 80
Canary current, 8
Cape Cod 66
Cape Hatteras, 8, 55
Carbon Cycle, 13
Carson, Rachel, 2
Cattail, narrow-leaved, *49*, 85, 108
Cellular respiration, 12
Cenchrus tribuloides, 51, *61*
Chamaecyparis thyoides, *39*, 53
Chenopodium album, 50
Chenopodium rubrum, 50, *71*, 86
Chesapeake Bay, 20
Chicken claws, 109
Chlorinated hydrocarbons, 16
Chondrus crispus, *106*
Ciguatera poisoning, 97
Cladonia cristatella, 36
Cladonia rangiferiana, 36
Cloverleaf fern, 37
Club fungi, 32
Coast-blite, 50, *71*, 86
Cocaine, 94
Coca plant, 94
Cocklebur, *52*
Coffea arabica, 94
Common reed, 72, *73*, 87
Conifer, 39
Continental shelf, 6
Conyza canadensis, *103*, 136
Cord-grass: Pacific, 73; smooth, *67*
Corema conradii, 77

Cottonwood, 64
Cotula coronopifolia, 73
Cowlicks, 69
Cross-pollination, 45
Croton, *62*, 63
Croton punctatus, *62*, 63
Crown-of-thorns, 100
Crustose lichen, *35*
Cuticle, 5
Cyanobacteria, 28
Cyperacea, 138
Cyperus papyrus, 138
Cypress, bald, *39*

DDT, 16
Delaware Bay, 20
Diatomaceous earth, 11
Diatoms, *10*
Dichotomous key, 117
Dicotyledons, 42
Dinoflagellates, 75
Dioecious plants, 49
Distichlis spicata, 50, *68*, 69
Ditch grass, 22
Doldrums, 7
Dulse, 20, *30*, 106
Dune sandspur, 51, *61*
Dusty miller, 59, 78

Earthstar, *34*
Eastern blue-eyed grass, *78*
Ecological seasons: autumn, 77; spring, 76; summer, 76; winter, 76
Eelgrass, *21*, 44
Erythroxylon coca, 94
Estuaries, 20
Euphorbia polygonifolia, 56, *57*, 82, 99
Euphorbia pulcherima, 100
Euphorbia splendens, 100
European sea-rocket, 58, 80
Euthamia tenuifolia, 82, *83*

Evening-primrose, seaside, 46, *62*, 63
Everglades, 24

F.A.A., 127
False heather, 59, *60*, 78
Fern: bracken, 38, *98;* cloverleaf, 37; lace-frond grape, *38;* mosquito, 36, *37;* netted chain, 38
Flat-topped goldenrod, 82, *83*
Fleabane, salt-marsh, *89*, 136
Florida current, 8
Foliose lichen, *35*, 36
Fruticose lichen, *35*, 36
Fucus vesiculosus, 19, 29
Fuller's earth, 11
Fungi: bracket, 32, *33;* club, 32; sac, 31, *33*

Geastrum saccatum, 34
Gibbons, Euell, 105
Gill mushroom, *32*, 33
Glasswort: perennial, *68*, 69, 88; slender, 109
Glycoside, 94
Goldenrod: flat-topped, 82, *83;* seaside, 46, *60*, 61, 73, 84, 103, 136
Golden heather, 79
Golden hedge-hyssop, 81
Golden pert, 81
Gratiola aurea, 81
Green algae, 28
Groundsel tree, *61*, 71
Gulf stream, 8
Gulfweed, *9*, 29, 34
Gymnosperms, 39

Heather: false, 59, *60*, 78; golden, 79
Heavy metals, 15
Hedge-hyssop: golden, 81; yellow, *81*
Herbarium, 118
Hibiscus moscheutos, 48, 90

High marsh, 69
Holdfasts, 19
Honckenya peploides, 59, *60,* 80, 108
Horse-latitudes, 8
Horse-mint, *102*, 136
Horseweed, *103*, 136
Hudsonia ericoides, 79
Hudsonia tomentosa, 59, *60*, 78
Hydrocotyle bonariensis, 62, 63

Ilex vomitoria, 62, 63
Intertidal zone, 18
Ipomoea sagittata, 63
Irish moss, *106*
Iva frutescens, 50, *68*, 69, 88
Iva imbricata, 59, 87

Japanese rose, 47
Jointweed, *83*
Juncaceae, 138
Juncus, 138

Kelp, 29; broad-leaved, *19*, 75; winged, *107*
Knotweed, seaside, *80*
Kombu, 107
Krill, 11

Labrador current, 10
Laccaria trullisata, 33
Lace-frond grape fern, *38*
Laguncularia racemosa, 25
Laminaria agardhii, 20
Laminaria saccharina, 19, 75
Lance-leaved milkweed, 46, *47*
Lathyrus maritimus, 50, *59*, 79, 109, 111
Laver: purple, 75; red, 75, 106
Law of the sea, 17
Leaf print, 131

Leaf skeleton, 131
Lichen: crustose, *35;* foliose, *35*, 36; fruticose, *35*, 36; reindeer, 36
Life history, 133
Limonium carolinianum, 72, 87, 103, 136
Linnaeus, Carl, 110
Live oak, *64*, 94
Loblolly pine, *99*
Low marsh, 67
Lycoperdon perlatum, 34

Mangrove: black, *25;* red, *24;* white, *25*
Maram grass, 55
Maritime marsh elder, 89
Marsh elder, 50, *68*, 69, 88
Marsh spurry, pink, 85
Milkweed, lance-leaved, 46, *47*
Minamata, 15
Mold, slime, 34
Monarda punctata, *102*, 136
Monera, 4
Monocotyledon, 42
Monoecious plants, 49
Morning glory, arrowleaf, 63
Mosquito fern, 36, *37*
Moss, Irish, *106*
Mushroom, gill, *32*, 33
Myrica cerifera, 63
Myrica pensylvanica, 36, 50, *63*

Narrow-leaved cattail, *49*, 85, 108
Neap tides, 7
Netted chain fern, 38
Neurotoxic poisoning. *See* Shellfish poisoning
Nicotiana tabacum, 94
Nicotine, 94
Nor'easter, 8
North Atlantic current, 8
North equatorial current, 8
NSP. *See* Shellfish poisoning

Oak: bear, 64, 94; live, *64*, 94; scrub, 64, *94*
Ocean currents, 8
Oenothera humifusa, 46, *62*, 63
Old man's beard, 36
Opuntia humifusa, 44, *45*, 79, 103
Orache, seabeach, *84*

Pacific cord-grass, 73
Pacific silverweed, 73
Palmaria palmata, 20, *30*, 106
Pamlico sound, 20
Panne, 69
Papyrus plant, 138
Paradichlorobenzene, 125
Paralytic shellfish poisoning. *See* Shellfish poisoning
PCBs, 16
PDB, 125
Pennywort, *62*, 63
Phragmites australis, 72, 73, 87
Phytoplankton, 10
Pickerelweed, 73
Piedmont, 54
Pigeon foot, 109
Pine: loblolly, *99;* pitch, 39, *64*, 94; pond, 39, *53*, 94
Pinus rigida, 39, *64*, 94
Pinus serotina, 39, *53*, 94
Pinus taeda, 99
Plankton, 10
Pitch pine, 39, *64*, 94
Plantago maritima, 73, 86
Plantain, seaside, *73*, 86
Plant dormancy, 75
Plant press, 119
Pluchea odorata, *89*, 136
Pneumatophores, 26
Poaceae, 137
Poinsettia, 100
Poison ivy, *99*
Poison oak, 99
Poison sumac, 99

Polychlorinated biphenyls, 14
Polygonella articulata, 83
Polygonum glaucum, 80
Pond pine, 39, *53*, 94
Porphyra leucostica, 75, 106
Porphyra umbilicalis, 75
Potentilla pacifica, 73
Prevailing westerlies, 8
Prickly pear cactus, 44, *45*, 79, 103
Protoctista, 34
Prunus maritima, 36, *51*
PSP, 96
Pteridium aquilinum, 38, *98*
Puffball, 34
Purple laver, 75
Purslane, sea, *58*

Quercus ilicifolia, 64, *94*
Quercus virginiana, *64*, 94

Ragweed, common, 95
Red algae, 30
Red laver, 75, 106
Red mangrove, *24*
Red Sea, 28
Red tide, 96
Reed, common, 72, 73, 87
Rhizophora mangle, 24
Rockweed, 19, *29*
Rosa rugosa, 47
Rose: Japanese, *47*; salt-spray, 47
Rose-mallow, swamp, *48*, 90
Rough-stemmed boletus, 33
Ruppia maritima, 22
Rush family, 138
Rye, 133

Sac fungi, *31*, 33
Salicornia europaea, 69, *70*, 88, 108, 110
Salicornia pacifica, 73

Salicornia virginica, *68*, 69, 88
Salsola kali, 53, 69, *81*
Salt-marsh afalinus, 52, 69, *70*, 87
Salt-marsh sand-spurry, 69, *70*, 85
Salt-meadow grass, *68*, 69
Salt-spray rose, 47
Saltwort, 53, 69, *81*
Samphire, 69, *70*, 88, 108
Sand-spurry: salt-marsh, 69, *70*, 85; sticky, 73
Sandspur, dune, 51, *61*
Sandwort, sea-beach, 59, *60*, 80, 108
Sandy laccaria, 33
Sargasso sea, 9
Sargassum filipendula, 29
Sargassum fluitans, *9*, 29, 44
Scrub oak, *94*
Seabeach orache, *84*
Sea-beach sandwort, 59, *60*, 80, 108
Sea blite, 73; southern, 69, 90, 108; white, 90
Sea chickweed, 80
Sea-elder, *59*, 89
Sea hollyhock, 91
Sea lavender, 72, 87, 103, 138
Sea lettuce, 19, 28, *29*, 75, 107
Sea-oats, 55, *56*
Sea oxeye, 72, 84
Sea-purslane, *58*, 80
Sea-rocket, *58*, 80; European, 80
Seaside-goldenrod, 46, *60*, 61, 73, 84, 103, 136
Seaside knotweed, *80*
Seaside spurge, 56, *57*, 82, 99
Seaweeds, 19
Secale cereale, 133
Sedge family, 138
Self-incompatibility, 48
Self-pollination, 49
Sesuvium maritimum, 58
Sesuvium portulacasrtum, 59
Shellfish poisoning, 96
Silica gel, 136

Silverweed, pacific, 73
Sisyrinchium fuscatum, 78
Slime molds, 34
Smooth cord-grass, 67
Solar wind, 1
Solidago sempervirens, 46, *60*, 6l, 73, 84, 103, 136
Southern sea-blite, 50, 69, *70*, 90, 108
Spartina alterniflora, 67
Spartina foliosa, 73
Spartina patens, *68*, 69
Spearscale, 49, *71*, 86, 104, 108
Species Plantarum (Linnaeus), 110
Spergularia macrotheca, 73
Spergularia marina, 69, *70*, 85
Spike-grass, 50, *68*, 69
Spring tides, 6
Sticky sand-spurry, 73
Suaeda californica, 73
Suaeda linearis, 50, 69, *70*, 90, 108
Suaeda maritima, 90
Sushi, 107
Swamp rose-mallow, *48*, 90
Sweet tangle, 19, 75

Tall wormwood, *53*, 84
Taxodium distichum, *39*
Thalassia testudinum, *22*, 49
Toadstool, 97
Toxicodendron pubescens, 99
Toxicodendron radicans, *99*
Toxicodendron vernix, 99

Trade winds, 7
Turtlegrass, 22
Typha angustifolia, *49*, 85, 108, 111

Ulva lactuca, 19, 28, *29*, 75, 107
Uniola paniculata, 55, *56*
United States Geological Survey (USGS), 115
Usnea, 36
Usnic acid, 98

Vasculum, 114

Wax myrtle, 63
White mangrove, *25*
Widgeon grass, 22
Winged kelp, *107*
Woodwardia areolata, 38
Wormwood: beach, 59, *60*, 78; tall, 52, 84

Xanthium strumarium, 52

Yaupon, *62*, 63
Yellow hedge-hyssop, *81*

Zooplankton, 10
Zostera marina, 21